Burpee
SEED STARTER

Burpee
SEED STARTER

A Guide to Growing Flower, Vegetable, and Herb Seeds Indoors and Outdoors

Maureen Heffernan

BURPEE
A Burpee Book

MACMILLAN • USA

We are grateful to the following companies for providing
the photography for this book:

W. Atlee Burpee Co.
Waller Flower Seed Cp.
Kieft Bloemzaden B.V.
K. SAHIN, Zaden B.V.

Sakata Seed America, Inc.
Pan American Seed Co.
Ernest Benary Seed Growers Ltd.
Ball Seed Co.
All America Selections

MACMILLAN
A Simon & Schuster Macmillan Company
1633 Broadway
New York, NY 10019

MACMILLAN is a registered trademark of Macmillan, Inc.

BURPEE is a registered trademark of W. Atlee Burpee & Company

Library of Congress Cataloging-in-Publication Data

Heffernan, Maureen.
Burpee seed starter: a guide to growing flower, vegetable, and
herb seeds indoors and outdoors / Maureen Heffernan.

p. cm.
Includes bibliographical reference and index.
ISBN: 0-02-861002-4 (pbk)
1. Plant propagation. 2. Seeds. 3. Gardening
 I. W. Atlee Burpee Company. II. Title.
SB119.H44 1997
635'.0431—dc21

96-49783
CIP

Manufactured in the United States of America

10 9 8 7 6 5 4 3 2 1

Book design by Rachael McBrearty

Book production by March Tenth, Inc.

Illustration of cold frame in Chapter 3 courtesy of Elayne Sears.

This book is dedicated to the memory of my grandfather, Frank Brown, who provided the family farm on which I was raised, and to the memory of my father, Dennis Heffernan, who also had a great love and appreciation for farms and gardens. This book is also dedicated to my aunts, Nancy McCauley and Nora Heffernan, whose gardens and lives continue to inspire me.

CONTENTS

CHAPTER 1

The Basics of Seed Starting
1

CHAPTER 2

Indoor Seed Sowing
11

CHAPTER 3

Outdoor Seed Sowing
43

ACKNOWLEDGMENTS

Many thanks to John Meils who helped with the initial coordination and editing of this book. Special thanks to Barbara Berger, Laurie Barnett, Sharon Lee, and Rebecca Payes of Macmillan for the final editing of the book. Thank you to Susan Johnston Carlson for her fine illustrations and Cheila Klaber for her technical editing.

I would also like to thank George Ball who asked me to write this book and his staff at Burpee, especially Barbara Wolverton and Elda Malgieri, who readily provided photos and information.

Finally, a special acknowledgment of thanks to the Cleveland Botanical Garden's Eleanor Squire Library. It houses one of the country's finest horticultural libraries with a helpful and knowledgeable staff.

FOREWORD

Where does seed come from?

How many of us can say more than just "from plants." Seeds are so small, inexpensive, and ubiquitous that they are often taken for granted. However, seeds are the very sparks of civilization—the source for sustaining human and animal life. The seed packet that you hold in your hand contains a fascinating and complex creation. Let me tell you a little more about what seeds truly are by telling the story of just one—the 'African Queen' impatiens seed.

Imagine, for a moment, that you are a green and growing plant, a wild impatiens living on the shaded bank of a river in Africa. Your flesh is green and your "food" consists of light, air, water, and soil. You spread your leaves to catch the sun and push your roots into the loose soil to stretch out and grow. Everything is fine until you get a little older and while you want to grow even more, alas, there is no room. The only way you can continue to expand is by reproducing yourself, so you begin a complex and ingenious series of changes.

Because you are growing in deep shade, you develop bright yellow flowers with a sweet nectar that attracts butterflies and moths to visit regularly to drink. While visiting your flower, these insects pick up a biochemical material called pollen on their legs. When they fly away to drink from another flower, they carry your pollen, which then gets attached to the new flower's pollen receptor. This "crossing" allows you and your neighbors to combine your genes. The pollen serves to fertilize an "egg" inside the flower. The results of this complex, mysterious, and seemingly random fertilization process are seeds.

You produce this seed along your green arms and fingers, as do your neighbors in your many tightly coiled wombs or "seed pods," and just when the seed is ripe, you release the coil and explode hundreds of seeds into the river. Your tiny seeds float with the current until some of them edge toward the riverbank and attach themselves to the mud there, where they wait for the right temperature and light conditions to sprout and grow.

The river's edge fills with your family of plants. And these plants in turn produce seed that may someday attach itself to the side of a fallen tree or perhaps a boat moored in the harbor. These seeds may journey down the river and wash up on yet another riverbank several hundred or even thousands of miles away to grow and start a new colony. This pattern has been repeated through the millennia by thousands of different kinds of plants resulting in the beautiful cloak of green that covers the earth.

A horticulturist traveling the world to discover new garden plants spots these plants growing and collects some of their seeds. He grows out the seed and selects the healthiest and most beautiful plants to keep. After more crossing, selections, and rigorous testing (which can take years), he finally introduces 'African Queen', the first yellow impatiens plant, to American gardeners.

In the world of horticulture, growers re-create this natural process of crossing plants to produce seed under more controlled situations.

Commercial seed production is done in areas of the world with even, mild temperatures like southern Europe, the Middle East, the highlands of Central America (especially Costa Rica), and coastal California. The seeds in your packets may have come from one of these areas or even Taiwan, Chile, or India.
When the seeds are ripe, but before the plant naturally releases it, the growers and farmers collect it. This is done by hand for very tiny seeds like impatiens and by machines with, for example, a marigold vacuum harvester that gently pulls ripe seed upward from the plants.

The harvested seed is then organized by type of plant, bulk packed in bags, and numbered by "lot." A lot number is a coded description of the day and year of harvest, location of farm, and place on the farm (for example, row, field, or greenhouse bench). This way any problem or question about the seed you order can be traced and,

we hope, addressed by pinpointing the cause of where and how it was grown. Look for the lot number on your next package of seeds. It appears on every package of Burpee seeds. Once the seed is labeled, the employees at the Burpee headquarters in Warminster, Pennsylvania, take up the immense task of packaging the many tons of seeds into individual packets. These seed packets are then described in our annual Burpee catalog, which is mailed to our customers.

Impatiens 'African Queen'

In a way, our customers can be likened to the birds and butterflies that are attracted to brightly colored flowers or fruits in nature. After receiving our annual catalog, gardeners select the plants they want to "drink from" and send in their orders. The orders are filled and shipped; and once again, seeds are released into the world.

Burpee is proud of the fact that the seeds we release are of extremely high quality. In fact, we still maintain the same rigorous standards for germination and genetic purity as set down 120 years ago by Washington Atlee Burpee, the founder of our company.

To learn to successfully grow the seeds we offer in our catalog as well as many other seeds, we proudly present this seed-starting guide for American gardeners.

This is a thoroughly researched, easy-to-use, comprehensive guide to growing flower, vegetable, and herb seed varieties. We believe the reader will find *Burpee Seed Starter* to be an engrossing and informative book, destined to be an indispensable feature on the gardening bookshelf.

George Ball, Jr.

INTRODUCTION

This book was sown many years ago when my grandmother gave me 50 cents by to buy some garden seeds at the Perry Coal & Feed Store in Perry, Ohio. My grandmother and I were in the store buying chicken feed when I spotted the new spring seed racks on display. As enticing as a candy counter, I was captured by the rich fantasy world and delight these packets promised. With just a few packets of seed, I could create my own little flower world; grow juicy, delicious sweet corn and watermelons; and best of all, have huge phantom pumpkins for Halloween.

So at eight years old with my parent's help, I set to work hoeing up a patch of ground on our farm to plant those seeds. I was skeptically optimistic that I could actually get these seeds to grow. I watered them and checked on them several times a day, but day after day, nothing. Three, four, five days went by . . . still nothing. Would they never grow? About the sixth day, when I was close to losing faith, a faint trail of bright green marigold seedlings appeared. I gave a shout and went running to tell my family. They all came out, and we stood around them marveling at this miracle.

Twenty-seven years and hundreds of seed packets later, I am still astonished when I see seedlings poke through the soil. In fact, I have even more appreciation for seed as I gain knowledge about their fascinating and complex workings.

This book is all about sharing that enthusiasm and knowledge for growing plants from seed.

Many people today are turning to gardening as a way to relax, exercise, be creative, and connect with nature. It is said that gardening has become the number one hobby in America. If one gets even a little serious about gardening, sooner or later, it leads to the "need for seed." But why the need for seed when garden centers are bursting with plants?

Well, for example, the person who started gardening by planting out a few sweet basil plants will soon have the need for seed when he also wants to grow cinnamon and lemon basil but can't find those plants at a garden center. And the gardener who saw a gorgeous display of 'African Queen' impatiens and 'Spring Sherbert' poppies at a flower show and wants to grow them in her garden will need seed when she realizes they are usually available only from seed. Or what about growing some of your neighbor's unique yellow pear tomato plants that he proudly says were passed down to him from his grandfather?

Unless you can start seed, you'll hit a horticultural glass ceiling and be limited to the variety of plants sold in garden centers or mail-order catalogs. Seed starting enables you to grow hundreds of more kinds of plants, including new hybrids, unique heirlooms, and other rare or unusual annual and perennial plants that are available only from seed.

As more and more people get into gardening, they also soon learn that gardening "ain't cheap"—a few flats of this and a few pots of that can add up to a frightening sum in no time. Seed starting is the cheapest method of plant propagation and is, therefore, a great way to save a considerable amount of money especially if you purchase a lot of annual bedding plants each year.

For example, compare the cost of a packet of zinnia seeds with a flat of 36 zinnia plants. A packet of zinnia 'Starlight' with 100 seeds costs about $2.00, but a flat of 36 of these plants will cost anywhere from $12.00 to $16.00, if you can find them. Even if you use only half of the seeds, you still save at least $11.00. As an added bonus, by saving the leftover seed to use the following year, you are essentially saving about $22.00 by buying one packet of seed. Now multiply this amount by the number of annuals you grow every year and you can see how seed sowing may just help you retire earlier.

By starting your own seeds you can also ensure that you will have healthy plants for your garden. Plants purchased at garden centers can sometimes be pot-bound, leggy, or otherwise of poor quality.

To take advantage of all of these benefits and others, this book contains complete information on how to start seeds indoors and outdoors. If you already have experience in growing plants from seed, this book should serve as a useful reference for providing specific germination requirements for all of the most commonly grown herbaceous garden plants.

The individual plant profiles also include information on the basic growing needs of each plant, including soil and light needs. Frost charts, a temperature zone map, and vegetable planting charts will further help you plan your gardening.

Just remember that starting seeds, like horticulture itself, is an art and science. There is no one absolute method for mastering either. Keep trying and experiment until you find some methods that work best for you. If you have problems or questions, take advantage of the various information sources listed in the appendices.
I guarantee that you will find growing plants from seed to be an interesting, rewarding experience and an invaluable skill to master.

Maureen Heffernan

3 New
Ivy Leaved Nasturtiums
Introduced in 1904 by
W. Atlee Burpee & Co.
Philadelphia
Copyrighted 1903
504
Painted from Nature

THE BASICS OF
SEED STARTING

TIMING

In seed starting as in life, timing is everything. Hurry a project, and it may turn out weak and disappointing; wait too long, and opportunities vanish. Luckily, knowing when to start seed is a relatively exact science.

Be careful of the powerful late winter impulse fueled by cabin fever to begin seed prematurely. When a balmy day arrives in late January—right after seed catalogs have arrived—it's tempting to begin sowing seed, since spring must be just around the corner. But starting seed too early is a waste of time, money, and materials. Seed started too early will often result in weak, leggy, and root-bound plants that are useless at planting time.

Most annual flowers need to be started 6 to 8 weeks before the last frost. Most annual vegetable seeds can be sown in the garden directly or started indoors 2 to 8 weeks before the last frost.

Most perennial seeds need to be started 8 to 12 weeks before the last frost. Keep in mind that a number of perennial seeds need a cold treatment before actually being sown indoors. Depending upon the seed, cold treatment should begin several days to several months before sowing dates. Cold treatment involves exposing seed to temperatures of 32 to 40°F.

Some seeds that need cold treatment before sowing include columbine, Iris species, and daylilies. Why do seeds need a cold period? Think about a plant's natural cycle. Seeds ripen in the summer or fall and drop to the ground. The seeds are exposed to cold temperatures until the spring when warmth and moisture stimulate its germination. Through evolution some plant seeds have developed a need to be exposed to cold temperatures before they'll germinate.

It is thought that this is Nature's way of protecting seeds from germinating too early in the season in which they are produced. If seeds that ripen in early fall germinated at this time, the resulting tender seedlings would find it difficult to withstand harsh winter conditions. This too-early germination is prevented because the seeds are genetically programmed not to germinate until they have been exposed to a lengthy cold period.

FROST-FREE DATE

What exactly is the frost-free date? The frost-free date is the approximate day in the spring when there should be no more hard or killing frosts for the season in a certain region. Therefore, temperatures after the frost-free date, even at night, should remain above 35°F.

If you don't know when the frost date is in your region, consult the frost-free chart dates. These dates factor in many years of weather patterns and represent an average frost-free date in your area. The exact date will, of course, vary from year to year, by several days or even by several weeks.

The frost-free date is an important consideration when planning to sow seed. For example, if you happen to live in Cleveland, Ohio, the average frost-free date is April 30. The first step is to get a bright marker and circle April 30 on your garden-planning calendar. This date will be your baseline for determining when you should begin starting most kinds of seed indoors. It will also let you know when you can begin sowing annual seed directly outdoors.

Let's say a gardener in Cleveland wants to grow some New Guinea impatiens seeds. She consults a planting chart and notes that impatiens need to be started 8 to 10 weeks before the last frost. Counting back from April 30, our gardener notes that this seed should be started between February 19 and March 5.

Eager to get the garden planted as soon as possible, this gardener now checks to see how soon she can direct sow the seeds that don't need to be started indoors. She plans to spend the first weekend after April 30, when all danger of frost has passed, sowing out this type of annual flower and vegetable seed.

The inevitable exception to this planning process is the unpredictable nature of weather. While the frost-free date may arrive as anticipated, the soil may be too waterlogged to plant out if spring rains were especially heavy. Or a cold snap may descend, with lower than average temperatures resulting in a late frost danger. Regardless of the situation, a gardener must be patient and wait until soil and planting conditions are ready.

INDOORS VERSUS OUTDOOR SOWING

Some seeds, like poppy, corn, and peas, need to be directly sown into the garden because they do not transplant well and their root systems are easily damaged. Other seeds, like impatiens and tomatoes, should be started indoors to get a jump on the growing season. And many seeds, like basil, marigolds, and cucumbers, can be sown either indoors or outdoors with equally successful results.

Seed that must be started indoors is done so for several reasons. Generally, such seed needs a longer growing season to form mature flowers or fruit in time to be fully enjoyed or harvested. Let's stay

TIP:

It's a good idea to use a calendar for planning your garden and to mark the general time periods that are commonly recommended for starting seed. These periods are usually 10 to 12 weeks, 8 to 10 weeks, 6 to 8 weeks, and 2 to 4 weeks before the frost-free date. Don't forget to mark the frost-free date as well. Write down the names of seeds on the dates that they ideally should be started. To know when to start specific kinds of seeds, consult the individual plant profiles presented later in this book.

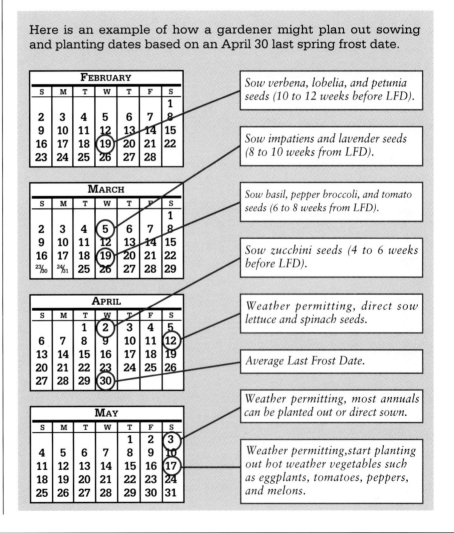

Here is an example of how a gardener might plan out sowing and planting dates based on an April 30 last spring frost date.

Sow verbena, lobelia, and petunia seeds (10 to 12 weeks before LFD).

Sow impatiens and lavender seeds (8 to 10 weeks from LFD).

Sow basil, pepper broccoli, and tomato seeds (6 to 8 weeks from LFD).

Sow zucchini seeds (4 to 6 weeks before LFD).

Weather permitting, direct sow lettuce and spinach seeds.

Average Last Frost Date.

Weather permitting, most annuals can be planted out or direct sown.

Weather permitting, start planting out hot weather vegetables such as eggplants, tomatoes, peppers, and melons.

Caution: While it may be tempting to sow all your garden seed at one time and consolidate your efforts, don't start seeds together that differ widely. The reason why seed-starting times vary is because some plants grow more slowly and/or need a longer development time before they can be planted out in the garden. Also, many seeds need a long indoor growing period before they start to develop flowers in time for spring or summer or to develop fruit before the first fall frost. Impatiens, rosemary, lavender, tomatoes, peppers, and eggplant all need a long indoor growing period before being planted in the garden.

with the impatiens example: If these seeds are sown directly into the soil after the frost-free date in most areas, they won't begin to bloom until late in the summer. By starting the seed indoors, the plants have added time to develop flowers early in the season.

Seeds that should be sown directly into the soil usually germinate quickly and grow fast enough to skip advanced indoor sowing. Plants that do poorly after being transplanted, like nasturtiums and beans, are usually sown directly.

You should sow annual seeds directly outdoors after the last frost date. Annual seeds thrive in soil around 60°F. If the soil is cooler than that, the seeds may rot before they germinate—especially if the soil is wet from heavy rainfalls.

There are exceptions to sowing seeds directly in the soil after the last frost date. Some seeds, including cornflower, gaillardia, cosmos, cleome, calendula, and sweet pea, can be sown directly outdoors as soon as the soil can be worked in the spring. Complete details for a wide range of annuals can be found in the individual plant profiles.

There are some plants that prefer to be sown outdoors. In cooler growing climates with shorter growing seasons, however, these plants should be started indoors to ensure adequate development before frost. It is important to sow these seeds into individual peat pots or pop-up pellets. When the seedlings are ready to be planted, the whole peat pot or pellet can be directly transplanted into the garden, as both will disintegrate in the soil and allow the roots to freely grow. Soil blocks are also effective and can be directly transplanted into the garden.

SEED PACKETS

Seed catalogs and package labels give quite a bit of information about how and when to start seeds, how to transplant seedlings, and what growing conditions are needed. You'll also find a description of the mature plant. Here are some typical entries for flowers and vegetables from the annual Burpee catalog and information from several seed packets.

BUYING SEED

It's best to buy and/or collect fresh seed each year. The fresher the seed, the better the germination results.

When buying seed, I look for hybrid seeds. These seeds result from the crossing of two different parent varieties. Each parent is selected for certain desirable characteristics. Hybrid seeds have

Seed packets from reputable companies furnish a great deal of information about how to grow the seed. Always check for the date that the seed was packaged to determine freshness. Only buy packets with the current growing season's date on it. For example, seed for the 1997 growing season is dated for the 1997 crop year.

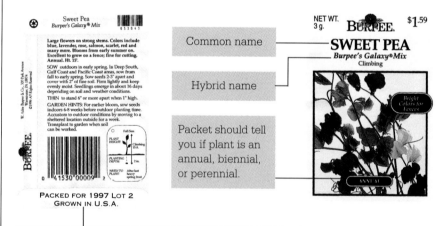

Date, lot number, country of origin
Most seed packets are labeled with a date, lot number, and country of origin. This seed was grown and harvested in the U.S.A. in 1997. The lot number identifies the specific field or greenhouse within the U.S.A. where the seed was grown. If any problems arise with the seed, like poor germination, a lot number generally enables a company to trace back to where the seed was grown to investigate any problems that may have occurred in growing and harvesting that particular seed variety. Seed was packaged in 1997. Unless stored in cool dry conditions, germination rate of older seed is significantly reduced.

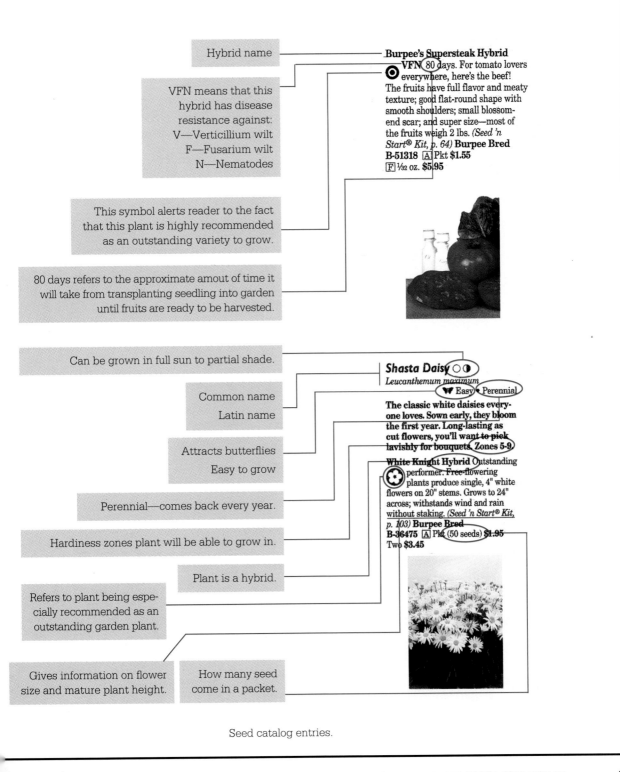

Hybrid name

VFN means that this hybrid has disease resistance against:
V—Verticillium wilt
F—Fusarium wilt
N—Nematodes

This symbol alerts reader to the fact that this plant is highly recommended as an outstanding variety to grow.

80 days refers to the approximate amout of time it will take from transplanting seedling into garden until fruits are ready to be harvested.

Burpee's Supersteak Hybrid
VFN 80 days. For tomato lovers everywhere, here's the beef! The fruits have full flavor and meaty texture; good flat-round shape with smooth shoulders; small blossom-end scar; and super size—most of the fruits weigh 2 lbs. *(Seed 'n Start® Kit, p. 64)* **Burpee Bred B-51318** Ⓐ Pkt $1.55 Ⓕ ¹⁄₃₂ oz. **$5.95**

Can be grown in full sun to partial shade.

Common name
Latin name

Attracts butterflies
Easy to grow

Perennial—comes back every year.

Hardiness zones plant will be able to grow in.

Plant is a hybrid.

Refers to plant being especially recommended as an outstanding garden plant.

Gives information on flower size and mature plant height.

How many seed come in a packet.

Shasta Daisy ○ ◐
Leucanthemum maximum
❀ Easy • Perennial
The classic white daisies everyone loves. Sown early, they bloom the first year. Long-lasting as cut flowers, you'll want to pick lavishly for bouquets. Zones 5-9.
White Knight Hybrid Outstanding performer. Free-flowering plants produce single, 4" white flowers on 20" stems. Grows to 24" across; withstands wind and rain without staking. *(Seed 'n Start® Kit, p. 103)* **Burpee Bred B-36475** Ⓐ Pkt (50 seeds) $1.95 Two $3.45

Seed catalog entries.

been bred to have a number of superior characteristics, including increased growing vigor, better disease resistance, and a number of ornamental or edible characteristics. Hybrid seeds often cost a bit more, but the increased vigor, health, and productivity of these plants are well worth the extra expense.

Nonhybrid seed (also called open-pollinated) are also worthwhile and have a place in every garden. Heirloom seeds, which have received increased attention in recent years, are simply nonhybrid or open-pollinated seed that is from old varieties handed down through generations of gardeners. One advantage these seeds have over hybrids is that nonhybrid seed can be saved from year to year. Hybrid seed should not be saved, since the second generation of plants will not "come true," that is, will not have the same characteristics as the parent seed.

If you want to save leftover hybrid or nonhybrid seeds for the following year, store them in their packets or a labeled envelope placed in a clean, dry, sealed glass jar. Place the jar in a cool, dry place. Before sealing the jar, fill it with uncooked rice, which acts as a moisture absorbent.

Seed Viability

If seed has been saved from the previous season(s) or you've been given seed from an unknown source, you may want to perform a seed viability test a few weeks before sowing. Seed more than a year old and/or that has been improperly stored from previous seasons may not germinate well or at all, since the "energy" stored in seeds loses strength over time. Some seeds may not produce the number of plants needed for your garden plan. Furthermore, older seed can sometimes produce weaker and less-vigorous plants.

Here is a simple method to test seed viability. In fact, it is one of the methods that the Burpee Seed Company uses to test its seed lots before packaging.

Take two paper towels and lay them evenly on top of each other. Thoroughly moisten them with warm water and sprinkle 10 to 20 seeds in the middle of the towels. Wrap up the towels, starting from one corner, like a crescent roll. Fold over at each end

TIP:
If you are unsure of the quality of the seeds you have purchased or have been given, it's always smart to test for viability. It's important to buy seed from a reputable company and only purchase seed dated for the year you want to use it.

about 3 inches. Place the towel in a glass containing about 1 inch of water—enough to keep the towel moist but not so much that the seeds are standing in water. Cover the glass with a clear plastic bag to keep in moisture, and put it in a warm place.

If you want to test viability for seeds that need light to germinate, like impatiens or begonias, place the seed on thoroughly moistened towels but do not fold the towels. Place them in a clear plastic bag and place in a warm spot. Open the bag and mist as needed to keep seeds moist.

Use the plant portraits in this book to note the average germination time for the seeds that you are testing. After the specified period of time, unwrap the towel and see if any seeds have sprouted. If they haven't sprouted yet, rewrap the seeds and wait a few more days. If any seeds have sprouted, count them and divide by the total number of seeds that were placed in the towel. This figure will give you a fairly accurate germination rate for the total batch of seeds.

For example, if you want to test the viability of pumpkin seeds that have been saved for 2 years, place 10 in the moistened towel. After 14 days (the average germination time), unwrap the towel and count how many seeds have germinated. If 3 of the seeds have germinated, you'll have about a 30 percent germination rate (3 ÷ 10 = 0.30) if these seeds are planted out. If you use these seeds, plant extra seeds to compensate for the low germination rate. Purchasing new seed ensures a better germination rate and stronger plants.

In addition to the seed that comes in seed packets, there are other specially prepared or packaged seeds that are time-saving and efficient.

PELLETED SEED

Pelleted seed is coated with a biodegradable material. The coating serves two purposes: It may contain nutrients or a fungicide to protect the seed from disease, and it may be used to enlarge seed size for easier handling, allowing for more even sowing. Pelleted seed can be spaced out at the mature plant distances so the need for thinning is eliminated. Be sure to thoroughly water these seeds after sowing. A good soaking will help break down the coating materi-

TIP:
Pelleted seeds and seed tapes are useful planting aids for folks who have arthritis problems as well as for any physically or visually impaired individual who cannot grasp or see the seeds well enough to sow them evenly. Seed tapes reduces the need to bend over and thin out seedlings.

als, which will permit the seed to more quickly absorb water. Note that for pelleted seed, the type of coating is indicated on the label or package.

SEED TAPES

Another way of packaging seed for garden use is by embedding them in seed tape. Seed tapes are especially popular for use with vegetable seeds such as carrots, herbs such as basil, and flowers such as cosmos and marigolds—seeds that can be directly sown in the garden. The seeds are evenly spaced down the length of a narrow piece of tissuelike paper that will dissolve after planting. Seed tapes should be laid out in regular planting drills and lightly covered with soil. Make sure the entire tape is covered with soil; any uncovered tape will act as a wick and dry out the tape and seeds.

Seed tape is particularly effective with small, hard-to-handle seeds that require even sowing in straight rows. Since seeds are evenly sown down the length of the tape, it reduces and often eliminates the need for thinning out seedlings and gives young plants plenty of space for early and healthy development. The tapes usually come in two lengths: 7 feet by ½ inch and 15 feet by ½ inch. The 15-foot tape can be cut into two or more pieces and used for multiple row plantings or for a second sowing later in the season.

Seed tape.

INDOOR SEED SOWING

WATER

Water is the spark that ignites the germination process. Once sown, seeds and seedlings must never be allowed to dry out, even temporarily. Conversely, just as water can cause damage by flooding, the germination medium must never become too waterlogged, which can often kill seedlings. Most seedlings are damaged by being overwatered rather than underwatered. Overwatering tends to make soil temperatures colder (especially in the spring when seeds need warmth) and reduces the amount of oxygen around the roots. The roots need oxygen to help take up water and nutrients; without it, the roots are susceptible to disease, especially damping off—the dreaded Grim Reaper of seed sowing. Strategies to avoid damping off will be discussed later in this chapter.

Horticulture is a balance of science and art. Keeping the germination medium at the correct moisture level is part science and part art. After some seed-starting experience, achieving this important balance will become easier.

TEMPERATURE

The optimal temperature for seed germination varies by plant species. However, almost all seeds need a warm soil temperature of 70 to 75°F to germinate. Seeds thrive in consistently warm temperatures while germinating. If left in cold soils, seeds may rot before they have a chance to germinate. To provide even heat, many gardeners use heating mats or heating cables. These devices not only provide steady heat but allow the temperature to be adjusted to the specific needs of a particular seed.

A number of perennial seeds need a "cold temperature treatment" before being sown. To subject seeds to a cold treatment, simply expose them to 32 to 40°F temperatures. The treatment helps break seed dormancy. Some seeds, like flowering cabbage, need just a few days of cold treatment; other seeds, like daylilies, need several weeks of cold treatment; and other seeds, like cardinal flower, need several months of a cold treatment. Once the cold treatment has been completed, the seeds should be exposed to warm temperatures for germination.

AIR

Air must be able to penetrate into the potting medium so seedlings can develop a strong, healthy root system. If the soil mixture is too heavy or overwatered, air can't penetrate; and the roots will rot. To avoid rot and promote air circulation, it's important to use a substance like perlite in the germination mixture to improve drainage and add air pockets in the medium.

LIGHT

Some seeds need light to germinate and other seeds need darkness. Indoor sown seeds that need light can be sown onto the soil surface and just lightly pressed into contact with the medium. Because the seeds aren't covered by medium, you'll need to cover the potting container with clear plastic or glass, which will hold in moisture.

Seeds that need darkness to germinate are covered with potting medium. Large seeds can be covered with ½ inch of medium, which will block out all light. Medium seeds are usually covered with about ¼ inch of medium. Very small seeds are sown onto the surface and lightly pressed onto the surface. Then the potting container is covered with black plastic, several layers of newspaper, or other opaque covering to keep light out.

The plant portraits in this book indicate the light requirements for seed germination. The seed packets should also indicate germination requirements.

Once any seed germinates, the resulting seedling needs light to develop. If seedlings are placed in a sunny southern exposure, natural light should be adequate unless the weather is excessively overcast. If space or natural light is limited, greenhouse or fluorescent lights should be used. Set up the lighting tubes a few inches above the seedlings and raise them as the seedlings grow. Various lighting setups are discussed later in this chapter.

Here are the materials you'll need to start seeds:
- Containers
- Potting mixture
- Heating mats or cables (optional but recommended)
- Fluorescent lighting tubes (optional, but usually necessary)

- Clear plastic or glass covering for seeds that need light to germinate
- Newspapers, burlap, or black plastic for seeds that need darkness to germinate
- Bulb water sprayers
- Labels and waterproof marker or grease pencil

CONTAINERS

Seed-starting containers can range from specially designed germination trays or pots to yogurt and orange juice containers.

Collecting germination containers is a marvelous opportunity to show off your creative recycling skills. Seasoned seed starters enjoy outdoing each other in discovering household items that make good germination pots.

While there is almost no end to usable containers, select ones that are at least 3 inches deep. If a container is too shallow, the germination mix dries out very quickly. Although your containers do not have to have drainage holes, I recommend using ones with drainage so that you can "water from the bottom," which helps prevent waterlogged soil. To water from the bottom, set the pots in a tray that has about ½ inch of water in it. If your containers don't have drainage holes (or holes can't be poked in the bottoms), be sure to place extra sphagnum peat moss, gravel, or polystyrene packing peanuts in the bottom of each pot.

If you're using standard plastic germination flats, a general rule of thumb is that a 5½ by 7½-inch flat will hold about 100 seedlings from large seeds, 200 seedlings from medium seeds, and 300 seedlings from very fine seeds. Always sow about twice as many seeds as you'll need plants, as many seedlings are lost to thinning and transplanting, and some seeds just won't germinate.

Peat pots and pop-up pellets (3 inches wide and deep) are best for plants that don't like to be transplanted. Peat pots are biodegradable and can be planted directly in the garden when the seedling is ready. The natural fibers of the container allow water and air to penetrate into the roots. Peat pots will disintegrate a short time after planting.

Pop-up pellets are compressed circular peat disks that, when moistened, expand to about 2 inches high. Seeds can be sown directly in them, and the entire unit can be transplanted into the garden when the seedling is ready. Kord Fiber Packs are similar to peat pots. They can be used for starting seeds and allow bottom watering.

If you don't use peat containers for seeds that are averse to transplanting, make sure the containers are large enough to allow the seedlings to develop without becoming pot-bound before they are ready to be planted outside. A plant becomes pot-bound when its container becomes too small for its roots to grow freely outward; instead, the roots begin to circle around the inside of the container and get thick and gnarled. Pot-bound plants often don't transplant and grow as well as plants whose roots have more space for development.

Soil block molds are used to shape moistened potting mix into individual soil containers. To make such a container, enough water is added to a potting mixture to form a thick slurrylike consistency, which is molded by the soil blocks. The soil blocks, as well as other germination containers, peat pots, and pop-up pellets, are available from most garden centers, seed catalogs, and garden-supply catalogs.

Burpee's Seed 'n Start and Window Sill Greenhouse containers.

POTTING MIXTURE

After optimal environmental conditions (water, temperature, light, and air), the potting mix is the most important factor in successful seed starting. Seed-starting media must be lightweight, well-aerated, and sterile (that is, free of disease). The mix should be kept evenly moist so it doesn't form a crusty top that is difficult for seedlings to punch through. Don't use soil from your yard. The chances are pretty good that it's too heavy and will cause disease problems for the seedlings. Instead, purchase or make your own sterile soilless seed-starting mixture.

Soil block molds.

Plastic germination trays allow for bottom watering.

Fertl-Cubes are made from a blend of mosses, vermiculite, and nutrients. The mix is pressed and dried into 1-inch cubes. Seedlings need to be transplanted into larger containers for growing on.

Peat pots and pop-up pellets are the easiest ways to sow seeds and plant seedlings. The containers are made from peat and wood fibers and contain a soluble fertilizer for healthy seedlings. Pots biodegrade when planted outdoors.

A PotMaker turns strips of newspaper into a plant pot. It's a great way to recycle newspaper.

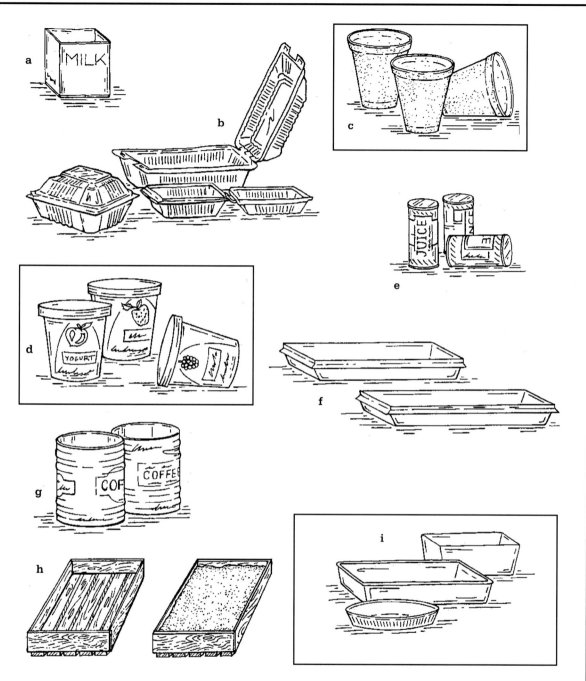

Here are some commonly available containers that can be recycled for seed sow-
ing: **a.** milk cartons (cut in half), **b.** plastic salad containers with lids, **c.** polystyrene
cups, **d.** yogurt containers, **e.** frozen orange juice containers, **f.** plastic water trays
(available from wallpaper stores), **g.** coffee cans, **h.** wooden crates, and **i.** 3-inch
aluminum baking pans.

Burpee's Seed-Starting Formula can be used for both flowers and vegetables. A soilless mixture provides optimal conditions for germinating seeds and growing on seedlings, for up to 8 weeks. If used for growing on after 8 weeks, fertilize the seedlings with a dilute liquid fertilizer.

It's easy and economical to make your own potting mix. Mixtures can be made from perlite, vermiculite, and sphagnum peat moss—all are available from any garden center.

Perlite is a natural substance, although it looks like an environmentalist's worst nightmare—a combination of plastic and polystyrene. In truth, it's a type of volcanic rock that has been superheated and puffed up to a lightweight pill size. Vermiculite is also made from a natural substance—mica. Sphagnum peat moss is composed of partially decomposed bog plants. Its main characteristic is its great water-holding capacity; yet it provides excellent drainage. Perlite is used to create good aeration and drainage in the potting mix. Vermiculite absorbs water well and keeps the mixture evenly moist. Sphagnum peat moss keeps the potting mix moist and aerated.

There is no one absolute best seed-starting mix. Experiment and come up with your own favorite mixture. Here are three recipes to try.

Mix #1. This is a good all-around recipe for starting most seeds. Mix equal parts (by volume, not weight) of perlite, vermiculite, and milled sphagnum peat moss. If you can't make drainage holes in a container, place a layer of sphagnum peat moss on the bottom before filling with the mixture.

Mix #2. For seeds sown into containers that will be directly transplanted into the garden (peat pots, pop-up pellets, etc.), use a recipe with a little nutrient boost. Here are two options:

1. Mix equal parts of a potting soil mixture, perlite, and vermiculite together.

2. The Cornell Peat-Lite Mixture was developed at Cornell University (Ithaca, New York) for starting seed. Although it can purchased ready-made, here's how to make it yourself:

- 1 bushel shredded sphagnum peat moss
- 1 bushel gardening vermiculite
- 4 level tablespoons ammonium nitrate (nitrogen source)
- 2 level tablespoons powdered superphosphate (phosphorus source)
- 10 level tablespoons dolomitic limestone

Mix #3. One part fine peat to one part horticultural sand.

Before filling the containers, lightly yet thoroughly moisten the potting mixture with warm water. Warm water is absorbed faster into the mixture than cold water. Soilless potting medium can be difficult to moisten thoroughly and evenly. I like to use a large metal wash tub to mix and moisten the ingredients. (These tubs are also handy for holding extra potting mixture until it's ready to be used.) Incorporate a little bit of water at a time into the mixture until entire mixture is evenly moist. For best results, use your hands to vigorously mix the soil. Before adding peat moss to the mix, submerge it in warm water for a few minutes and wring out any excess water before adding it to the potting medium. Peat moss can act as a drying agent in the mixture

NOTE: *If you aren't starting a lot of seeds, you'll probably want to purchase a ready-made bag of starting mix rather than buying all the individual ingredients.*

unless it has retained enough water to keep it moist. After moistening the mix, fill the containers up to about ½ inch from their tops. Lightly press the potting mix down and add a bit more mixture, if necessary, so the medium remains ½ inch from the top.

Metal tubs are handy for mixing and moistening potting media.

SOIL-WARMING DEVICES: ELECTRIC HEATING MATS

Most garden seeds need steady temperatures of 70 to 75°F (night and day) for optimal germination. Some seeds even need germination temperatures as high as 85°F.

It's often difficult, if not impossible, to find one spot where the temperatures are warm enough and steady enough 24 hours a day.

NOTE: *After germination when seedlings have emerged, night temperatures can be lowered by about 10°F. If you are using a heating mat, simply unplug the unit or lower the temperature setting at night. Once seedlings have been transplanted into larger containers, they will not need extra heat.*

You can try placing containers on top of the refrigerator or radiator or near a wood stove or heat vent. However, the latter three options can also quickly dry out the potting mix. For best results, invest in a heating mat designed for starting seeds. These units are placed under the pots and provide bottom heat directly at soil level, ensuring even and optimal soil warmth for seeds and

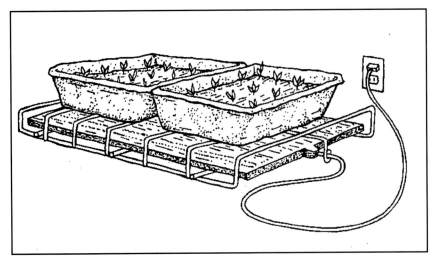

Electra Grow Mat units are safe and easy to use. Trays or flats are placed on a wire rack just above a rubber heating mat. This device warms the soil 15 to 20°F above the surrounding air temperature.

seedlings. If germination containers are placed near a window for light in late winter or early spring, heating units are critical, since temperatures can be much cooler near the window than in other parts of the room. Remember that if soil is too cold, germination and seedling growth can stall or fail altogether.

Heating units can also save you money on your heating bill. Instead of turning up the central heat, use a heating mat to deliver heat only where it is needed: to the potting mix. It is the potting mix temperature—not the air temperature—that is key.

Lights

Once seedlings have emerged, they need light for proper development. If adequate light is not provided, growth will be weak, spindly, yellowish, and leggy. (Leggy stems are stretched out between branches, which makes the stem and branches weak and the plants unattractive.) Leggy seedlings are starving for light and will literally stretch to find it.

Seedlings need 12 to 14 hours of light per day for good development. A sunny southern to southwestern exposure is usually adequate during the spring. A heated or, in warmer climates, an unheated greenhouse will provide enough natural light. If you don't have these conditions or have filled up all the sunny places in your home or greenhouse, you'll need to provide seedlings with some artificial light.

Many beginning seed starters are intimidated by setting up artifical lighting, yet it is very easy to do. It involves using standard 40-watt cool-white or warm-white fluorescent light tubes and rigging them up so they can be adjusted to remain 2 inches directly above the seedlings, even as the plants grow. The more expensive wide-spectrum plant lights are not necessary for growing seedlings. Wide-spectrum lights are used to trigger flowering in plants produced by commercial propagators.

Use the longest tube possible to fit your growing space. Light at the ends of the tube is a little weaker, so the longer the tube, the more effective light area you'll have. Tubes should be spaced no more than 4 to 6 inches apart, if you're using more than one tube.

Light Carts

If you're handy with tools, try building a light cart. Investing in a commercial grow light cart can also be worthwhile. Light carts are self-sufficient, space-saving, plant-production powerhouses. Fitted with heating mats, light carts can even turn your basement into prime growing space. In fact, the money saved in the long run by starting seeds instead of purchasing plants will pay for the unit in no time. These units can also be used for propagating cuttings and growing houseplants year-round.

Grow light units range from tabletop designs to multitiered carts. There are units for every budget and growing need. You can make your own or purchase units from the Burpee catalog, garden centers, and garden-supply catalogs. Featured here is the Burpee Glow 'n Grow Light Garden.

Burpee's Glow 'n Grow Light Garden is portable and has two large shelves that can hold up to 48 six packs or 72 three-inch pots. Two adjustable lighting tubes are affixed on 12-inch chains above each shelf. As the seedlings grow, the lights can be easily adjusted to remain 2 inches above them. The unit is 47 inches high, 49 inches long, and 19 inches deep.

Tabletop Plant Light.

LIGHT CARTS

This light cart is intended for use in growing plants indoors under fluorescent lights. It is assembled from 1½-inch PVC pipe and fittings and has plywood shelves. It can be used on a tabletop or countertop, or it can rest on a platform or concrete blocks on the floor. The dimensions are designed for maximum flexibility. The unit can be set up as illustrated with room on the bottom shelf for storage of pots, potting mix, and other material; or it can be inverted to make the bottom shelf the growing area, with storage above. The plywood shelves are each 24 by 48 inches, the size of one-quarter of a sheet of plywood. Use ½- or ¾-inch plywood or another shelving material of the appropriate size. Shelves should be painted or otherwise protected from water, to prevent warping and fungal growth.

The light fixtures are standard 4-foot fluorescent shop light fixtures, fitted with cool-white fluorescent tubes. Commercial-grade fixtures, while more expensive, are better quality, have a quieter ballast, operate better at a variety of temperatures, and come with a longer cord. They also have larger reflectors and provide more light to the plants. Two of this type fixture are adequate for the unit. If inexpensive shop lights are used, three fixtures are needed to provide the same amount of light. It will take more electricity to operate the budget lights than to run the commercial-grade fixtures. A modification in the attachment of the lights (suspension from cross boards, for example, rather than hanging them by a chain from the PVC pipe) is also necessary if budget light fixtures are used.

The unit is wired by attaching a commercial light strip (with an on-off switch) to one of the bars of the light cart with plastic cable ties. The light strip is plugged into a grounded timer, which is plugged into the wall outlet. A humidity tent can be constructed from sheet vinyl or cut from a clear vinyl shower curtain. It can be sewn to fit over the entire structure, or vinyl sheets can be attached to the PVC crosspieces or light reflectors with sticky backed hook and loop tape.

Jim Flint
NGA

If you want to save a lot of money, and are handy with tools, you can easily make your own light unit. The National Gardening Association (NGA) has developed a do-it-yourself plan to build a light cart for indoor seed starting. Plans for this unit were made in conjunction with the organization's national GrowLab education program, which brings light carts into classrooms to help children learn about plants and gardening. This plan was originally developed by Joe Premo and Phil Tennison and modified by Char Bezanson, a biology instructor at St. Olaf College (Northfield, Minnesota). For more information on the GrowLab program, contact the NGA at 180 Flynn Avenue, Burlington, VT 05401 (802-863-1308).

MATERIALS

½ sheet of ½-inch plywood, cut into two 24

by 48-inch pieces for shelves

45 feet of 1½-inch PVC pipe, cut into pieces as shown in the diagram

8 PVC elbows to fit 1½-inch pipe

16 PVC tees to fit 1½-inch pipe

One 6-ounce can PVC primer

One 6-ounce can PVC adhesive

2 heavy-duty 48-inch 40-watt fluorescent shop light fixtures, grounded for safety

Medium-weight chain and 8 S-hooks for hanging light fixtures

1 multi-outlet electrical strip with cord and on-off switch, grounded for safety

1 timer, grounded for safety

4 plastic cable ties

IMPORTANT!

When using artificial lights, remember to turn them off after the seedlings have received 12 to 14 hours of light in a 24-hour period. Plants also need darkness for proper development. Iturn off the lights at night sometime between 8 and 10 P.M. and turn them on again by at least 8 A.M. Using a light timer to turn the lights on and off automatically is the most reliable method.

DIRECTIONS FOR ASSEMBLY

Paint (optional)

Powerstrip (optional)

ENDS

1. Lay one end out on the floor or on a large table so that the are aligned as shown in the End View. pieces Place all the pieces together without gluing them.

2. Use a pencil to mark where each piece comes together with and overlaps its partner. Be sure the angles are correct; number each joint.

3. Starting with one end, paint primer on each piece where it will join together with another piece. The primer helps the glue adhere to the PVC pieces. Glue the end assemblies together. Put glue around the end of each piece (in a ½-inch-wide band) and join the pieces. As you reinsert the pieces and line up the marks, work quickly. The glue dries very fast (within 15 seconds) and will not come apart again. Since the crosspiece of each end (F-T-D-T-F) are all the same, you might want to assemble these first.

4. Repeat this process with the other end.

CROSSBARS

5. With one end on the floor, insert the crossbars and mark the pieces as before. Prime and glue.

6. Fit the other end onto the six vertical crossbars. Mark as before and prime the pieces. Because all six crossbars and the remaining end are glued at the same time, you'll need the help of an assistant.

FINISHING

1. Place the shelves on the supporting crossbar. If desired, you can paint the shelves with an acrylic paint or cover with vinyl, plastic trays, or other material.

2. Assemble and hang the light fixtures according to the manufacturer's instructions, using the chain and S-hooks. Remember that the lights will need to be moved up as the seedlings grow.

3. Strap the outlet strip to one of the pipes, using the plastic cable ties. Plug the cord into the timer and then into the wall outlet. Be sure that you bought grounded fixtures, strip, and timer, for safety. If desired, you can use a power strip for surge protection instead of a regular outlet strip. If so, follow the same assembly as for the outlet strip.

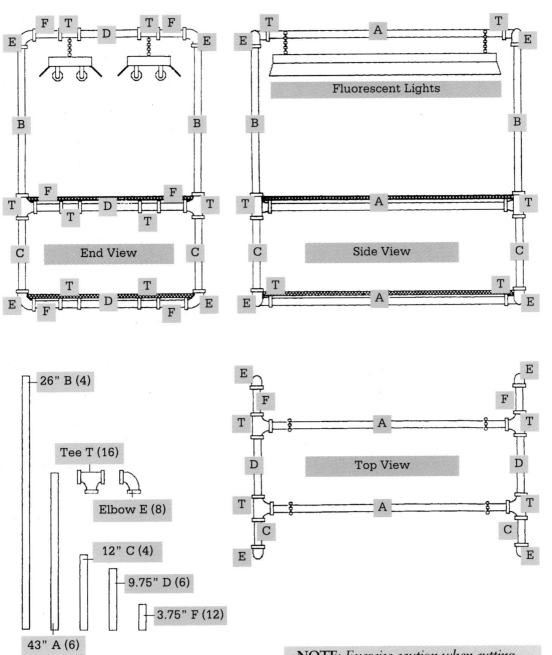

End View

Fluorescent Lights

Side View

Top View

26" B (4)

Tee T (16)

Elbow E (8)

12" C (4)

9.75" D (6)

3.75" F (12)

43" A (6)

NOTE: *Exercise caution when cutting the PVC pipe, working with the primer and glue, and wiring the electrical components. Wear safety glasses and a mask and work in a well-ventilated area.*

A bulb water sprayer is handy for watering delicate seedlings.

Bulb Water Sprayer

You'll need a bulb water sprayer to water seedlings. This device emits a light spray of water when the sides are squeezed. It's perfect for keeping the potting medium appropriately moist during germination and will prevent overwatering of young seedlings. Don't use a regular watering can or hose spray because each emits a heavy stream of water that will displace seeds, damage seedlings, and lead to waterlogged soil. Also, you can still bottom water seedlings as described earlier. Remember that seedlings don't need soil as moist as at the germination stage.

If you don't want to purchase a bulb sprayer, try using a mist spray bottle. However, it takes quite a bit of spraying with a mister to adequately water seeds and seedlings. A bulb sprayer is much more efficient and takes less time to get light, thorough water coverage.

Remember to mark seed name and sowng date on labels.

LABELS

Always remember to write the name of the seed and the sowing date on a label and place it on the germination container. It can be very difficult to remember all the different seedlings at planting time. To avoid getting the flowers in the vegetable bed and vice versa, remember to label your seeds. By marking the date of sowing you will know when to expect germination. Use plastic labels or wooden markers and write with a waterproof marker or grease pencil.

PRETREATMENTS

Before being sowed, some seeds require special treatment. The treatments include providing a cold treatment, scarifying, and soaking. (Individual pretreatment needs for seeds are noted in the plant profiles.)

COLD TREATMENT

Some seeds need a cold treatment to help them break down the outer seed coat so water is more easily absorbed. Cold treatment also helps break the inner chemical dormancy that must be overcome before the seeds can germinate. Cold treatment is required for many perennial seeds whose natural cycle in the wild includes a cold weather (winter) period before germination occurs in the spring.

Cold treatment can be given to seeds by various methods. Two of the easiest are as follows.

For the first method, seeds can be mixed with a small bit of moistened sphagnum peat moss (three times the volume of the seeds) and put in a tightly tied plastic bag, which is placed in a refrigerator at a temperature between 32° and 40°F.

If you have the space, seeds can be sown directly into germination containers or flats, covered with plastic wrap, and placed in the refrigerator. Keep the potting medium moist throughout the cold treatment period.

After the seeds have received the required minimum period of cold temperature, take them out of the refrigerator. If you've used a plastic bag, evenly scatter the contents of the bag onto the potting medium in a germination container. Don't try to separate the seeds from the sphagnum peat moss. If the seeds have already been sown into their germination containers, germinate them by providing their particular temperature and light requirements.

When scarifying seeds with a knife, be careful not to cut into the center of the seeds, which can harm them.

The second method is for seeds that need a lengthy cold treatment. Sow the seeds into pots in the fall and place them outdoors in a sheltered area over the winter. Put some plastic or a light covering of mulch over the potting medium.

SCARIFICATION

Some seeds, like nasturtium and moonflower, have tough seed coats. To speed germination, their seed coats can be scarred on a few sides with a nail file or sandpaper, or they can be lightly nicked in a few places with a sharp knife. Scarification allows water to quickly penetrate the seed embryo area, which prompts germination.

WATER SOAKING

For seeds like corn that take a long time to soak up water, germination can be speeded up with an intensive moisture treatment. This is achieved by soaking the seeds, usually overnight, in warm water just before they're sown.

GERMINATION: STEP BY STEP

Now that you've done any necessary pretreatment and have gathered your seed-starting equipment, you're ready to begin. Here's a step-by-step process for starting seeds indoors.

Materials Needed

Potting mix
Germination containers or pop-up pellets

Seeds
Bulb sprayer or mister
Plastic bag
Bottom-watering trays
Labels

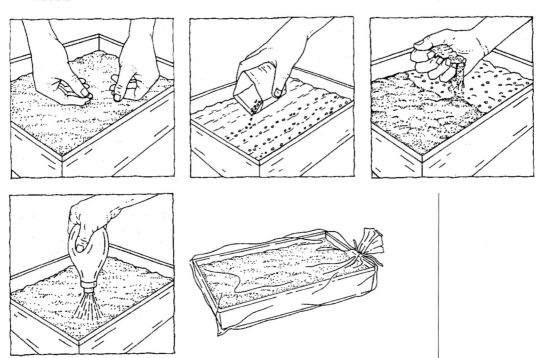

Step-by-Step Sowing Directions

1. Moisten the potting mix and fill the germination containers to ½ inch from the top. Lightly firm down medium. If you're using pop-up pellets, arrange them in the bottom of a tray and pour a few inches of water over them and let them fully expand.

2. Evenly scatter the seeds over the surface of the medium or plant them in shallow rows spaced about 1 inch apart for small seeds and 2 inches apart for large seeds. For individual pots, sow two to three seeds in each container. For pop-up pellets, place one seed in the center of the pellet and pinch the sides to cover the seed.

3. Lightly cover the seeds with potting mix to a depth of two to three times the size of the seed diameter. If the seeds need light to germinate, just press them onto the surface.

EXPERT TIP:
For seeds that are extremely small, like begonia, mix a little horticultural sand into the seed packet, which will allow you to sow the seeds more evenly.

Don't forget to label your containers; it's useful to mark the date of sowing and plant name.

4. Lightly water in the seeds with a bulb sprayer or mister, making sure the seeds are not displaced.

5. Label and cover the containers with clear or black plastic (depending on the seeds' need for light).

Provide Moist and Warm Conditions for Seeds

It's best to water the potting mix from the bottom up—this can be done only with containers that have bottom drainage holes. Set the containers in a plastic tray or liner and fill it with ½ to 1 inch of warm to tepid water. The potting mix will act as a wick to take up water from the bottom. This avoids the problem of displacing seed by overhead watering and decreases disease problems that occur when a seedling's foliage gets too wet.

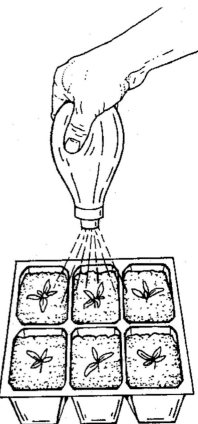

Be very gentle when watering seeds and seedlings. Use warm to tepid water in a bulb sprayer to water.

If you don't water from the bottom, then a bulb sprayer is the best method. Either method provides enough water so the potting mix is lightly and evenly moist, not heavy and waterlogged. Check the mix a few times a day to make sure it's not too dry or too wet. If the mix is too wet, empty the water from the bottom tray until the soil dries out a bit. Sometimes the symptoms of overwatering can look like underwatering: seedlings wilt and appear limp.

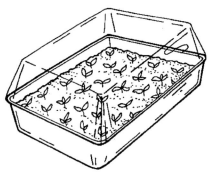

Seal in moisture and keep humidity high by covering the containers. However, remove the cover if condensation is noticeable.

To maintain high humidity levels during germination, cover the containers with a piece of glass, plastic bag, or other plastic covering. If water begins condensing on the cover, immediately remove it and allow to dry out a bit before placing the cover back over the container. If conditions are too damp and humid, you may experience damping off or other disease problems. Remove the cover once 50 to 70 percent of the seeds sown have germinated.

Since most seeds need temperatures of 70 to 75°F to germinate, place the containers in a warm site or use electric germination heating mats.

true leaves   Cotyledon

This is a seedling with its first set of true leaves compared to cotyledon leaves. Cotyledon leaves, also called seed leaves, are the first leaves to appear from a germinated seed.

Once the seeds have germinated, the seedlings need 12 to 14 hours of light per day. If you don't have an appropriate greenhouse or enough southern exposure, use fluorescent bulbs. See the discussion on lights, earlier in this chapter, for information on how to set up a lighting system.

After germination, the seedlings do not need temperatures as high to grow on as they did to germinate. Seedlings do not need electric heat mats. Day temperatures can be reduced to 60 to 65°F. Night temperatures can be

reduced by an additional 10°F. Reducing temperatures will make the seedlings sturdier and healthier.

THINNING

Thin seedlings at soil level so the remaining seedlings are not disturbed. Leave enough room for good air circulation and uncrowded growth.

When the seedlings have developed their first set of true leaves, they need to be thinned out and/or transplanted into larger containers so the remaining seedlings have adequate room to develop.

Seeds that were sown into individual cell packs or peat pots need to be thinned only. Leave one seedling in the container to grow on until it can be planted into the garden. Always select the largest, strongest, and healthiest seedling to remain. Thin out other seedlings by cutting then off at soil level with scissors or tweezers. This method will not disturb the root growth of the remaining seedling.

If seeds were sown into germination flats, thin so that the remaining seedlings have at least 1 to 2 inches between them for smaller plants and 3 to 4 inches for larger plants. If the flats are at least 3 inches deep, they can be used to grow on the seedlings. If not, transplant the seedlings into larger containers.

> **NOTE:** *Don't wait too long after the seedlings have developed their first set of true leaves to thin or transplant. If you do, seedling growth can be weak, stunted, and vulnerable to disease by overcrowded growing conditions. After transplanting, keep the soil moisture even. Give plants 12 to 14 hours of light. Transplants like a humidity level of between 50 and 70 percent. Mist seedlings once a day with a mister bottle.*

TRANSPLANTING

Seedlings are ready to be transplanted into larger growing containers once they reach about 2 inches in height. To help prevent root

1. Use a fork to lift the seedling from the flat.

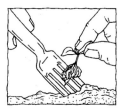

2. Hold the leaves in your hand and use the fork to hold up the roots.

3. Plant the seedling into the center of a 3 by 3-inch container.

4. Firm the soil around seedling.

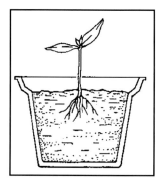

1. Seedling planted too shallow.

2. Seedling planted too deep.

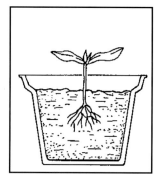

3. Seedling at correct depth.

damage, make the planting hole in the new container before lifting the seedling from its pot. The seedlings should be planted so they are slightly deeper than they were in the germination container. The transplanting container should be at least 3 by 3 inches.

To transplant, use a fork, spoon, pencil, plastic plant label, or a chopstick to gently lift up the seedlings from the potting mix. Use the tool to hold up the roots as you lift the seedling to transplant it. Once the roots have been lifted up, gently grasp seedling by the leaves and transplant into the prepared container. Be careful to hold the leaves lightly, since grasping them too hard can damage young plant cells. Seedling stems are especially prone to injury if held too tightly.

Lightly firm the potting mix around the stem base so the roots are in good contact with the soil. The same soilless mix used for germinating can be used for transplanting, or a regular potting soil mixture will suffice. Return the transplanted seedlings to their sunny location or back under the lights.

FERTILIZING

Once seedlings have been transplanted, give them a light application of fertilizer. Use a soluble, complete, and balanced fertilizer mixed at one-quarter strength solution and feed the seedlings every 7 to 10 days. A complete and balanced fertilizer means it contains nitrogen (N), phosphorus (P) and potassium (K) in even proportions. Look for a 5-5-5 fertilizer mix for seedlings. You won't need any extra fertilizer if you are using the Cornell Peat-Lite Mixture to grow on seedlings or transplants.

HARDENING OFF

Seedlings should not be planted outdoors until after the danger of frost and until they've been hardened off. Hardening off is the process of acclimating plants to harsh outdoor conditions. The reality of direct sun, drying winds, and cooler temperatures can wreak havoc on delicate seedlings. Seedlings should be hardened off 1 to 2 weeks before being planted outdoors.

Hardening off is usually done by exposing plants to increasing amounts of time outdoors. In fact, the hardening off process

Cold frames are an excellent way to harden off seedlings. Make sure the lid is lifted during the day so temperatures do not get too hot inside. Close the lid in late afternoon to trap the heat and keep the plants warmer at night.

should actually be started indoors 1 week before the seedlings go outdoors to harden off. During this time, cut back a bit on the watering and keep temperatures between 55 and 60°F if possible. After a week of indoor hardening, place plants outdoors in a sheltered, part-sun location for about half a day when day temperatures are above 50°F. Make sure plants do not dry out since water evaporates more rapidly outdoors.

Extend the time outdoors and the amount of sunlight plants receive each day for about 1 week. Plants can remain outdoors overnight after 4 to 5 days of hardening off, if there is no danger of frost. Keep an eye on the weather report and bring them back inside if a frost is forecast. It is important to harden off all the plants you have started indoors, even the plants grown from seed that could have been sown directly outdoors. Shade-loving plants don't need as much hardening off time as full-sun plants. After 1 week of hardening off, the plants are ready to go into the garden.

PLANTING OUT

After the danger of frost has passed, it's time to begin planting your garden—annuals should go in first. The exceptions are those vegetable plants like peppers and eggplants that like warmer soils. Wait and plant these hot-weather crops when the soil has warmed up in late spring or early summer. Remember to plant in well-prepared, well-drained soil. Incorporate compost and other organic matter into the soil to a depth of at least 12 inches for deeply rooted perennials and vegetable crops, especially root crops such as carrots, onion, and garlic.

Try to plant out on an overcast day since it lessens transplant shock even to hardened-off plants. Make planting holes in the soil two to three times bigger than the plant's root

Make a few slits in peat pots before planting them out.

EXPERT TIP:
Make sure that the tops of the peat pots and pop-up pellets are covered with soil. If part of the pot is above-ground, the moisture will wick up and evaporate out rather quickly, drying out the roots.

size. Gently take plant from its container. If the roots are girdling around the container, gently pull them out and straighten them so they'll grow out and not continue orbiting the root ball.

Plants should be placed slightly deeper in the ground than they were in their pots. Firmly press the soil down around the roots and stem. When finished with all the planting, give the plants a good soaking.

If transplanting a peat pot or pop-up pellet, plant the entire container directly into the soil. It's a good idea to take a sharp knife and make a few slits in the pots before planting them.

Check on your newly planted plants every day to see if they need more water. If some appear limp or wilted, check to see if the soil is in good contact with the roots. If not, the roots will quickly dry out. Mulch around the plants to retain soil moisture.

TROUBLESHOOTING SEED STARTING PROBLEMS

If you've carefully followed the seed-starting guidelines, but your seeds have failed to germinate, it may be due to one or more of the following problems.

Problem: The seed was too old and/or wasn't stored properly if it was saved from the year before.

Solution: Buy fresh seed each year or store leftover seed in a cool, dry place if it is to be used the next year. If you're unsure of seed vitality, do a germination test using wet paper towels (see page 8).

Problem: The soil was allowed to dry out.

Solution: Check the soil for moisture once or twice a day; don't place the germination containers near excessive heat or drafts. Cover the containers with glass or plastic.

Problem: Lack of light for seeds that need light to germinate.

Solution: Check before sowing if the seeds need light to germinate. If so, sow them on the potting mix surface and lightly press the seeds onto the surface.

Problem: Too much light for seeds that need darkness to germinate.

Solution: Check before sowing if the seeds need darkness to germinate. If so, cover the seeds with at least ¼ inch of potting mix and place black plastic or newspaper over the container until the seeds germinate.

Problem: Soil temperature was too cold. Containers were placed in a spot without adequate heat.

Solution: Do not put containers too near a window or on the floor where cooler air collects. Remember that most seeds need 70 to 75°F to germinate. Use an electric heating mat under the germination containers or place the containers near a heat source (radiator, heat vent, top of the refrigerator). Use warm to tepid water when watering; cold water can lower the soil temperature.

Problem: Soil temperature was too hot.

Solution: Make sure that the heating mats or other heating units are adjusted accordingly. Check the seeds before sowing for their optimal germination temperature. Sow seeds that need different temperatures in separate containers.

Problem: Seeds were planted too deep.

Solution: Check the proper sowing depth in the plant portraits or on the seed packet before sowing. The general rule of thumb is to bury seeds to a depth of two to three times their diameter. For very fine or tiny seeds, scatter them on the potting mix surface and lightly press them into contact with the mix.

Problem: The seeds were not in good contact with the potting mix.

Solution: Make sure the potting mix does not have too much air space. After filling the containers, lightly tamp down the mix, adding more as necessary. After sowing seeds, lightly press them into contact with the surface. If the seeds don't need light to germinate, then lightly tamp down the potting mix that covers the seeds.

Problem: The potting mix top was too crusty or the mix was too heavy or clayey.

Solution: Don't use garden soil, which can be too heavy and becomes crusty on top when it dries. Purchase a soilless mixture for seed starting or make your own by using equal parts vermiculite, perlite, and sphagnum peat moss. Make sure the medium is evenly moist before sowing the seeds.

Problem: The seeds were not given their required pretreatment.

Solution: Before sowing, check if seeds need a cold treatment, water treatment, scarification.

TROUBLESHOOTING ONCE SEEDS HAVE GERMINATED

Once damping off occurs, affected seedlings cannot be saved. To avoid the heartbreak of damping off, always use clean containers and a sterile soilless mix. Avoid oversowing, overwatering, and provide good air circulation.

DAMPING OFF

Damping off is the most dreaded and frustrating problem encountered when starting seeds indoors. Seeds germinate and seedlings appear healthy but suddenly wilt, topple over at soil level, and die for seemingly no reason. This sudden death is caused by fungi in the soil: Rhizoctonia, Pythium, Botrytis, or Phtyophthor. The fungus may have been present in the soil (especially if garden soil was used or a soilless mix was reused) or in the seeds themselves or spread by water. Here are the most common problems that lead to damping off and the precautions to take.

Problem: The soil is too wet.

Solution: Do not overwater the seedlings. Water from the bottom up, but don't let the containers sit in more than ½ inch of water. Use containers with drainage holes or add 1 inch or so of gravel or sphagnum peat moss to the bottom of all containers without drainage.

Problem: The seedlings were overcrowded.

Solution: Don't sow seeds too thickly. Thin out as soon as seedlings start to crowd each other or when they develop one set of true leaves. Give remaining seedlings 1 to 2 inches of space (more for larger plants).

Problem: Outdoor garden soil was used.

Solution: Always use a sterile soilless growing mixture. Do not reuse soilless mixes.

Problem: The germination containers were not thoroughly cleaned before use.

Solution: Wash all containers in soapy water with a 10:1 dilution of water to household bleach. Rinse well.

Problem: There was a lack of good air circulation.

Solution: Remove the plastic or glass covering after 50 to 70 percent of the seeds have germinated. Place the containers in a spot with good air circulation. Thin out seedlings so there is adequate air circulation around them. Some gardeners set a fan on low about 6 feet from the seeds for a few hours every day. (I like to place my light carts in my bedroom and dining room, which both have overhead fans to provide terrific air circulation for the seedlings.)

OTHER PROBLEMS. . . AND SOLUTIONS

Problem: Leggy seedlings.

Solution: Leggy seedlings have long, spindly stems with large gaps between the leaves. They are a result of one or more of the following conditions: inadequate light levels, excessively high temperatures, and overcrowding.

- Make sure the seedlings are getting 12 to 14 hours of bright natural or artificial light per day. Use 40-watt bulbs if using fluorescent lights.

- Make sure the lights are not more than 2 inches above the seedlings.

- Make sure the seeds are getting 6 to 8 hours of darkness per 24-hour period

- Lower growing on temperatures by 5 to 10°F during the day and 10°F at night from the original germination temperatures. Most seeds will grow on with daytime temperatures of 60 to 65°F and nighttime temperatures of 50 to 55°F.

- Thin out and/or transplant seedlings into deeper growing containers.

Leggy seedling versus a healthy seedling.

Problem: Leaf discolorations.

Solution: Discolored foliage usually indicates that the plants need more light and/or nutrients. If the true leaves (not the cotyledons) are turning yellow, it may indicate a nitrogen deficiency. If the leaves are turning a bronzy color or turning brown around the edges, it often indicates a potassium deficiency. And if leaves are reddish purple on the undersides, it may indicate a phosphorus deficiency.

Problem: Mold.

Solution: Scrape off the mold but be careful not to disturb the seedling's roots. Use only dilute, light fertilizer solutions and no more than once a week. Use containers with drainage holes, water from the bottom, and do not overwater.

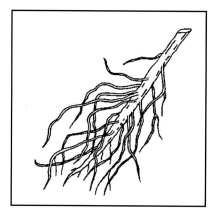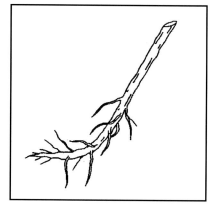

A strong, healthy root system compared to a weak, stunted root system.

Problem: Poor root growth.

Solution: Seedlings with poor root growth may be suffering from one or more of the following problems: poor drainage, low potting mix fertility, low soil temperatures, and lack of oxygen in soil.

Use containers and potting medium with good drainage; add perlite to soilless mixtures to increase drainage, and increase oxygen into the root area. Fertilize seedlings with a complete fertilizer mix. Add bottom heat or move the containers to a warmer growing on space.

Problem: If a seedling's leaves curl under even when exposed to bright light, the plant may be overfertilized.

Solution: Reduce fertilizing and repot the seedling into fresh potting mix.

Burpee's Offering
SEEDS, BULBS AND ROOTS FOR SUMMER AND FALL
1918

RUTA BAGA
IMPROVED PURPLE-TOP YELLOW

TURNIP
PURPLE-TOP
STRAP LEAVED

ROOT CROPS

EARLY
LONG SCARLET
SHORT TOP
RADISH

BURPEE'S
COLUMBIA BEET

CARROT
CHANTENAY

W. Atlee Burpee & Co.
Seed Growers   Philadelphia

OUTDOOR SEED SOWING

WHAT SEEDS GET STARTED OUTDOORS

Plants that don't transplant well and/or are fast growing can be sown directly outdoors. Greens and other vegetables are most often sown directly in the garden, but a number of annual and perennial flower seeds can also be directly sown.

Here is a list of some common garden plants that can be directly sown in the garden in early spring or after the last frost date.

VEGETABLES AND HERBS	FLOWERS
Basil	African daisy
Beans	Annual phlox
Borage	Baby's breath
Carrots	Bachelor's buttons
Chamomile	Bells of Ireland
Chervil	California poppy
Coriander	Chinese forget-me-not
Corn	Cosmos
Cress	Immortelle
Cucumbers	Larkspur
Dill	Love-in-a-mist
Fennel	Marguerite
Greens (mustard, collards etc.)	Marigolds
Lettuce	Mignonette
Melons	Nasturtium
Onions	Poppies
Parsley	Sunflower
Peas	Sweet alyssum
Pumpkins	Sweet pea
Radish	Tahoka daisy
Spinach	Zinnia

Sowing seeds successfully outdoors depends on three things: timing, soil preparation, and weather conditions.

TIMING

Before sowing seeds, know your average last frost date and check the recommended sowing times for each seed type. Many of the seeds listed here (for example, peas) can be directly sown in early spring as soon as the soil can be worked. However, tender annuals and vegetable seeds, like corn and melons, should not be sown directly until all danger of frost has passed and the soil has warmed up a bit more. If seeds are sown into soil that is too wet and cold, they may rot, germination can stall, and the seedlings may damp off.

If tender annuals are sown too early in the spring before the last frost-free date, they may germinate, but a frost may kill the seedlings.

Although a number of seeds can be sown directly, if you live in a northern climate with a short growing season, you'll need to start many of these seeds (such as tomatoes) indoors early so the plants will flower or bear fruit before the end of the season.

SOIL PREPARATION

Good soil preparation is the foundation for a healthy and productive garden. It is critical to prepare soil extremely well when sowing seeds directly outdoors. Your goal is to create the same kind of soil quality as you would use indoors: light, finely textured, and well drained.

The soil must be well drained. This is the key advice to remember about garden soil. If you are starting with heavy or clayey soil, you will need to work in several wheelbarrow loads of organic matter to improve it. However, your effort will be well rewarded, as you'll be able to grow healthy and vigorous plants.

Before working the soil, you need to determine if it is ready to be worked. If you begin preparing it when it is too dry or too wet, you may do damage by compacting it. Here's a simple method to determine if the soil is ready to be worked: pick up a handful of

soil and try to form a ball. If the soil crumbles, it's too dry; moisten it up a bit before working it. If the soil forms a wet, sticky ball and doesn't fall apart loosely when tapped, it's too wet; allow it to dry out before working it. Ideal soil forms a ball that will fall apart if tapped lightly. In early spring, if your soil has ice crystals in it, it is too cold.

To prepare the outdoor site, work the soil to a depth of 12 to 18 inches with a shovel or spade. Turn over the soil and incorporate at least 2 inches of finished compost, shredded leaf mold, or shredded peat moss. If the soil is very poor, add 4 to 6 inches of organic matter. Consider purchasing some good topsoil to improve the soil.

RAISED BEDS

Building raised beds is a good way of coping with heavy clay soils. Begin by breaking up soil clods with a short-tined rake or sharp spade. Remove any rocks, sticks, and twigs from the seed bed area. The soil should have a smooth, fine-textured, and even surface. If your soil is heavy and poorly drained, work in some sharp sand and consider installing some drainage pipes below the soil surface.

Try not to walk on the prepared soil in the seed bed areas. If you must traverse a wide area, lay down some boards or walk in garden pathways to avoid compacting the soil.

pH

Most plants prefer a slightly acidic soil, which means a pH of 6.0 to 6.9. If you don't know the pH of your soil, ask your local County Extension Office how you can get your soil professionally tested. You can also purchase soil-testing kits through the Burpee catalog or other garden-supply catalogs. If you need to adjust the pH of your soil to make it slightly more acidic, purchase some sulfur, aluminum sulfate, or iron sulfate from a garden center or nursery and follow the manufacturer's directions. To raise the pH (to make the soil more alkaline), incorporate crushed dolomitic limestone into the soil.

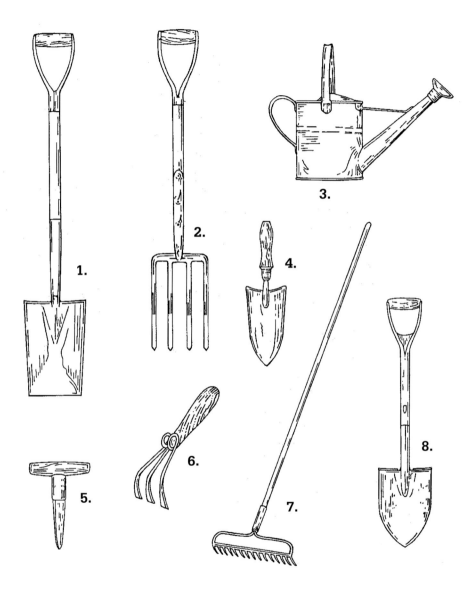

With all gardening jobs, using the correct tool helps gets the job done well. Good, sharp tools will save you time and energy. Invest in the best tools you can afford. Here are the tools you'll need to prepare soil, sow seeds, and transplant.

1. Standard shovel for digging and turning over soil.

2. D-handled spading fork for breaking up clods and aerating the soil.

3. Watering can with fine nozzle or hose with fine nozzle attachment.

4. Transplanting trowel.

5. Dibber, for making seed and transplanting holes and seed furrows through the soil.

6. Hand fork.

7. Bow rake for smoothing the soil surface and raking away debris.

8. D-handled spade for turning soil and breaking up clods.

How to Adjust the pH of Your Soil

Here is a general guide for changing your soil's pH value by 1 unit, to make it more acidic or more alkaline.

Till or dig the amendments into the top 4 inches of the soil 3 to 4 weeks before sowing seeds or planting. It's a good idea to do another soil test 1 month after making the adjustments to check that the pH is right. Over time, the soil will revert back to its natural acidity, so additional applications will be necessary to maintain the desired pH value.

How to Raise the pH Value by 1 Unit

Soil Type	Ground Limestone (pounds per 100 square feet)
Light sandy soil	3.0
Sandy loam	4.5
Loam	6.5
Silty loam and clayey loam	0 8

How to Lower the pH Value by 1 Unit

Soil Type	Sulfur (pounds per 100 square feet)	Aluminum Sulfate (pounds per 100 square feet)
Light sandy soil	0.5	2.5
Silty loam and clayey loam	2.0	6.0

Note: To raise or lower the pH by less than 1 unit, decrease the amount of amendment used by about half. If you have your soil professionally tested, you will receive advice about how to adjust the pH.

Sowing and Spacing

After the soil has been worked to a fine texture and smoothly raked, water the bed thoroughly for 1 to 2 hours with a sprinkler. This will allow the water to soak down 1 to 2 inches into the soil. Wait 1 day and then sow seeds into the moistened bed. If it rains heavily after the soil has been prepared, wait and sow seeds when the soil has partially dried out.

If planting a vegetable or cutting garden, you'll be sowing seeds in straight rows. Follow the directions on the seed package for sowing depth and spacing within and between rows. To make straight rows, use stakes and twine or string to measure and mark the rows.

Sow seeds evenly. Try not to allow clumps of seed to spill out in one place. If the seeds are large, handle them individually. If seeds are medium, carefully tap them out of the package as you go along the row. If seeds are extremely small, put a little sand in the seed packet and then carefully tap them out down the row.

Once the seeds are sown, go back and gently rake soil over the seeds and lightly tamp it. Seeds should be in direct contact with the soil. If the seeds need light to germinate, firmly press them into contact with the soil and sprinkle a light dusting of soil over them to keep them moist. Most seeds that need light to germinate are best starting indoors where moisture levels can be better controlled.

Remember to label the rows or areas where seeds have been planted; use waterproof markers on plastic, wood, or metal labels. This is particularly important if you're sowing seeds within established flower beds or borders so young seedlings are not trampled or weeded out by mistake.

Make your plant rows run north and south to maximize sunlight and decrease shading.

CREATING OUTDOOR SEED BEDS

If you have the space in your garden, consider creating a little plot just for starting seeds. A small area is especially useful for starting annual and perennial flowers. Sowing flower seeds into established borders is sometimes difficult because young seedlings must compete for light, water, and nutrients with the larger surrounding plants. Seedlings can be easily mistaken for weeds or grasses and pulled up or trampled if planted around other plants.

Seeds that need lengthy cold periods can be sown in this bed in the fall and transplanted to their garden site the following year.

Your seed bed should be situated in a sheltered spot with partial shade. Make sure the soil is well prepared for good seed starting; it should be finely textured, light, moist, and well drained.

If you are sowing seeds in the summer when sun and heat are relatively intense, cover the seeds with a shade cloth until they germinate or apply a light cover of straw or mulch to help retain moisture. Another good cover is moistened burlap sacks placed over the seed bed. Remove the cover as the seedlings develop.

Keep weeds away from seedlings, but be careful not to disturb the seedlings' roots when pulling nearby weeds.

WATERING

After the seeds are sown, they need to be "watered in." Use a bulb sprayer, watering can, or hose with an attachment that delivers a fine mist of water. If the water stream is too heavy, it will displace the seeds that you have just sown. A hard stream can uncover the seeds and leave them vulnerable to drying out. Check the seed beds for moisture every day, and never let the soil dry out. Keep it lightly and evenly moist at all times during germination and after the seedlings emerge.

After the seedlings have developed into larger plants, you should water the soil deeply about once a week. Most plants need about 1 inch of water per week. If you don't get enough rainfall, irrigate the plants for 1 to 2 hours each week. Plants develop better root systems if they are given long soakings instead of frequent light waterings.

THINNING

Once seedlings develop one set of true leaves or reach 1 to 2 inches tall, they need to be thinned out. Follow the seed packages' instructions or check the plant profiles in this book for optimal plant spacing. If you water the beds lightly before you start thinning, it will be much easier to pull up the unwanted seedlings. When thinning, be careful not to disturb the remaining seedlings' roots. (When I'm thinning out plants, I place one finger on either side of a remaining seedling so it doesn't get disturbed by the thinning process.)

Thinning plants is not easy work. It takes a strong back, padded knees, and resolve (I feel a bit guilty wasting the thinned-out seedlings). Thinning allows the remaining seedlings to grow strong and healthy. If you can't bear to throw away the seedlings, pot them up and give them to gardening friends and neighbors. Some seedlings, like herbs and salad greens, can be used in salads; or recycle seedlings in the compost pile. As you gain more experience in sowing seed, you'll learn how to plant more evenly, making thinning out an easier job.

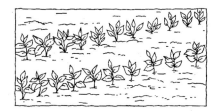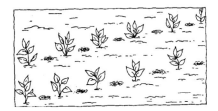

Once seedlings have developed true leaves or are 1 to 2 inches tall, thin them to required their spacing.

FERTILIZING SEEDLINGS

If your soil has been enriched with added compost, aged manure, and other organic matter once or twice a year, then lots of extra fertilizer is unnecessary, except for some heavy vegetable feeders such as corn and tomatoes. However, most soils and seedlings need a little fertilizer boost early in the growing season. There are exceptions, mainly some herbs, wildflowers, and ornamental plants that prefer less fertile soils. Check the growing requirements for your plants before fertilizing.

As a rule of thumb, apply a 5-5-5 or 5-10-5 fertilizer when seedlings reach 4 to 6 inches in height. Follow the manufacturer's directions for appropriate quantities. If your soil is very poor, apply a half-strength water-soluble fertilizer to young plants every week for 3 to 4 weeks after they have been planted. Some fertilizer can be worked into beds when the soil is prepared in the spring. Many annual flowers and vegetables also benefit from a midseason application of a light fertilizer.

Another approach to fertilizing involves using a slow-release fertilizer. Instead of applying an early and a midseason fertilizer, a slow-release pellet-type fertilizer, like Osmocote, can be applied around the base of plants in late spring or early summer. Slow-release fertilizers will last throughout the season, releasing fertilizer when it rains or the garden is irrigated. Even if you add extra fertilizer, it is important to improve soil naturally with organic matter. Organic matter not only increases fertility but also maintains good soil structure and drainage.

PROTECTING SEEDLINGS

The outdoors poses special hazards for developing seedlings, in addition to the problems encountered by seedlings started indoors. Be sure to guard against damping off: don't overwater, thin out the seedlings, and provide light, well-drained soil.

Slugs and cutworms pose special problems for young seedlings. Both pests are notably obnoxious and repulsive little creatures. To keep slugs at bay, spread some diatomaceous earth, crushed eggshells, or crushed seashells around the seedlings. Slugs will cut their soft bodies on the sharp edges and turn away. Cutworms mow

EXPERT TIP:
Follow the fertilizer manufacturer's directions carefully. If the label says apply 1 teaspoon, don't apply 1 tablespoon for good measure. Don't use powerful lawn-type fertilizers, like a 40-20-20, on your flower and vegetable gardens. Too much fertilizer, especially nitrogen, can harm plants and inhibit flowering and fruit development. Be careful not to spread fertilizer on stems or foliage, as it can burn delicate plant parts. Natural or organic fertilizers, like well-composted manure, can be incorporated into the planting hole or seed beds; but make sure they are well mixed with soil. While they are slower acting, organic fertilizers help improve soil structure and its natural fertility over time.

CROP ROTATION

Rotate the location of your edible plants each year. If you don't rotate crops, disease and pest problems will build up in the soil. You can also plant part of your vegetable garden with a green cover crop for one year (turn it under in the fall or following spring). The cover crop adds nutrients to the soil and builds up good soil structure. Call your County Extension Office for more information about sowing cover crops.

down seedlings at soil level. Plant collars made from metal cans (I like to use the larger tuna fish cans), cardboard rolls, plastic and cardboard milk cartons, and paper and polystyrene cups will control cutworms. Remove the bottoms of the containers and, if necessary, cut them so they're 2 to 3 inches in height. Place the collars around the base of the plants, leaving about 1 inch between the stem and collar. If the collar is more than 1 inch from the stem, cut through one side and staple the ends together to make a smaller diameter. The collars can be removed once the plants are well established, since cutworms prefer young, succulent growth.

If you see evidence of other pest damage, like chewed leaves, call your County Extension Office for prevention and control methods. Many botanical gardens and other horticultural organizations have free information services to answer your gardening questions and help with problems.

Animals such as rabbits and the cute but diabolical deer can wreak havoc on your plants. If rabbits are a big problem,

Protect seedlings from cutworm damage by placing collars around young seedlings.

cover seedlings with wire mesh. Deer damage be prevented only by installing fencing, either a standard fence at least 8 feet high or an electric fence around the garden perimeter.

There is a range of good books on controlling diseases, insects, pests, and animals in the garden. Two of my favorite books on this subject are *The Healthy Garden Handbook* by the editors of *Mother Earth News,* which has lots of practical advice for controlling pests without heavy applications of chemicals, and *Disease and Pests of Ornamental Plants* (5th ed.) by Pascal Pirone, which gives advice for both cultural and chemical control of diseases and pests.

TIPS FOR STARTING PERENNIAL SEED OUTDOORS

Many of the perennials in this book can be sown directly outdoors in the fall, spring, or summer. Ideally, perennial seed should be sown as nature intended, that is, when seed ripens and drops from the mother plant. For many garden perennials, this is usually midsummer to fall. You can allow plants to self-sow in the summer or fall and not disturb the seedlings until they are large enough to transplant the following spring or summer. If you don't want to transplant them anywhere else in the garden, thin out the seedlings so the remaining ones have enough space to grow properly . If you collect your own seeds, watch the plants carefully several weeks after flowering to see if their seeds have ripened. Cut off an old flower head and gently shake it to see if the seeds fall out. If so, cut off a few of the seed heads and collect the seeds to sow.

When sowing perennial seeds outdoors, it is best to sow them into a holding bed or in containers. Although perennial seed can be sown directly in beds or borders, there are some problems with this approach. It is easy to forget that seeds have been sown in those patches, and they can be dug up or planted over with other transplants by mistake. Young seedlings are frequently trampled or mistaken for weeds. The larger, established plants can shade the tiny seedlings, making them weak and leggy.

Here are some general guidelines for fall, spring, and summer sowing of perennial seeds.

FALL SOWING

If the perennial seeds need a long cold period, sow them into containers with a soilless mix in the fall. Water in the seeds, label the containers, and place them in a cold frame or other protected site in the garden. If the containers are not in a cold frame, add a layer or two of evergreen boughs, burlap, or other protective covering over the containers. The containers can also be sunk into the ground. An unheated garage or shed is a good site to overwinter seeds sown in containers. If stored indoors, bring the containers back outside to a sheltered site in the spring.

As the weather warms in the spring, make sure the seeds are kept evenly moist. When the seedlings begin to grow, thin them or transplant them into larger containers; half-gallon containers are a good size to grow on perennials. The plants can be transplanted into the garden in the summer or fall. If the seedlings are small and slow growing, wait until early fall or the following spring to transplant them into the garden.

SPRING SOWING

If your perennial seeds don't need a long cold period or have been given a cold treatment by refrigeration, they can be sown into an outdoor seed bed, sown into containers and kept outdoors, or sown directly in the garden,

Which method is best for you? Decide by a process of elimination: If there is no room to start the seeds indoors, then start them in an outdoor seed bed. If you don't have a spot for a seed bed, sow the seeds into containers and place them outside. If you don't want to transplant seedlings and if you have an uncrowded garden area, sow the seeds directly where they are to grow.

Whether seeds have been sown in the fall or early spring, once they begin to grow, they will need some extra help. Feed the seedlings a dilute, one-quarter-strength liquid fertilizer once every 7 to 10 days until they reach about 1 foot tall. Remember that most perennials will not bloom the first year when started from seed. However, some varieties like mealy cup sage, sunrise coreopsis, and Shasta daisy will be sown in the fall or early spring.

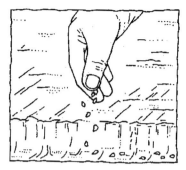
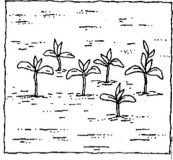

A nursery bed is ideal for sowing seeds and growing on seedlings until they are large enough to be planted out in the garden.

NOTE: *Seeds collected from hybrid and other cultivated varieties will not grow true to type. Either purchase hybrid perennial seed or propagate vegetatively (by cuttings or division) if you want more plants of a specific cultivar.*

SUMMER SOWING

Summer is a good time to sow freshly ripened perennial seed as well as packaged seed. Make sure that the seed bed area or containers are shaded, so the soil and seeds do not dry out. After germination, keep young seedlings moist and shaded. Even if they are sun-loving plants, harden off the seedlings by exposing them to increasing light levels before allowing full exposure to direct sunlight. Most perennial seed started in containers in early to midsummer should be transplanted into the garden the following spring. Some faster-growing perennials such as bee balm, black-eyed Susan, and silver dollar plant can be transplanted into the garden in the fall.

WHAT IF SEEDS DON'T GERMINATE OUTSIDE?

POSSIBLE PROBLEMS

- The soil was not prepared well enough; there are too many clumps or the clods were not broken up.
- The soil dried out and/or got too crusty or too hard for the seeds to break through.
- The seed was too old, not stored properly from the previous year, or was from an unknown source.
- The seed was not planted deep enough and so dried out before it could germinate.
- The seed was planted too deep, which often happens with very fine seed.
- The soil was too cold and wet and the seed rotted before it could germinate.
- The soil was too heavy, wet, and/or compacted, leading to a lack of oxygen.
- The seed was eaten by animals or birds (cover with a fine wire mesh or nylon netting until seedlings are more established).
- Heavy rains washed the seed away or heavy winds blew the seed away, which can happen with very light, fine seed.

Remember, it takes a little trial and error to get the hang of successfully sowing seed and growing on plants. Don't be discouraged if you don't achieve perfect results. Try to always learn from problems and prevent them from happening again. If you have a friend, family member, or neighbor who has a green thumb, invite him or her over and ask for troubleshooting advice for indoor and outdoor seed starting.

COLD FRAMES

Experienced gardeners consider the cold frame to be a garden tool as necessary as a good shovel or trowel. Cold frames are versatile and energy-efficient devices that help with many gardening tasks, from starting seeds and hardening off plants to forcing bulbs, over-wintering perennials, propagating cuttings, and growing hardy greens and vegetables in cool-weather months. They act as mini-

greenhouses and can be tucked away into a corner in your garden.

Cool weather, fast-growing vegetables and greens can be sown in the cold frame in early spring. Crops such as spinach, lettuce, turnips, radishes, and onions can be sown 4 to 6 weeks before the last frost. If sown at that time, you can harvest a crop of vegetables while other gardeners are just beginning to sow out their seeds.

Cold frames can help you get a jump on starting hardy annuals. Sow seeds into containers 2 to 4 weeks before the last frost and place the containers into the cold frame. Seedlings will get plenty of natural light, and daytime and nighttime temperatures will be just right. If you've run out of space to start seed indoors, cold frames give you more seed starting space.

To harden off indoor-grown seedlings, place them in the cold frame 1 to 2 weeks before they are to be planted outside. If the cold frame is in direct sun, cover the top with burlap or shade cloth for part of the daytime during the first several days of hardening off so the plants can get acclimated to stronger light levels. Plants can stay overnight in the cold frame as long as the lid is closed. However, if the temperature drops below 35°F while tender annuals are in the cold frame, make sure it's insulated.

Insulate the cold frame by doing one or more of the following: throw an old blanket or two over the entire structure; place at least four 1-gallon plastic jugs filled with warm water inside the cold frame (water holds heat); place bales of hay around the sides of the structure; and cover the plants inside with a layer of burlap, newspapers, or plastic. If you have installed soil-heating cables in the bottom of the frame, make sure they are turned on to keep temperatures above freezing.

By insulating the cold frame, many gardeners can enjoy fresh, tender greens almost year-round. In areas with more severe winters, this may not be possible. But there are many gardeners in Vermont, coastal Alaska, and other cold and snowy states who revel in the challenge to match wits with and defy winter by harvesting greens in the dead of January.

In late spring and early summer when the cold frame has been emptied of seedlings and hardened off transplants, it can be used throughout the growing season to start fresh perennial seeds as each species ripens.

The most important thing to remember with cold frames is to keep the lid open on warm days so plants don't dry out, burn up, or suffer from disease. The temperatures inside a closed cold frame can be 15°F or more higher than the outside temperature. If outdoor temperatures get over 45°F, prop up the lid a few inches; if temperatures get above 65°F, prop up the lid even more. Remember to close the lid again in the late afternoon to help trap the heat, so plants will be snug for the night.

In late summer or early fall, another crop of cool-weather vegetables, greens, and fast-growing herbs (for example, parsley and basil) can be sown. Once these are harvested in late fall, the cold frame swings into action for perennial seeds that need a lengthy cold treatment. Perennial seeds can be sown into containers and placed in the cold frame to overwinter. The cold frame will protect the seeds from the extremes of winter weather and speed up germination and growth the following spring. Don't forget to label your containers.

To germinate and grow any kind of perennial or annual seed in the cold frame, follow the same rules as you would for outdoor or indoor sowing: use a soilless mix, keep soil moist but not wet, and guard against damping off.

How to Build a Cold Frame

A cold frame can be as simple as a few bales of hay and an old window sash set up against a shed wall or as fancy as a redwood structure with automatic venting lids. It's up to the individual and depends on how much you can spend, how handy you are with tools, what materials you have lying around, and how cold it gets in your area. A cold frame is very easy to build.

The cold frames should be situated in a sunny location with a south to southeast orientation and protected from strong winds. Select a site with good water and soil drainage. Try to place the cold frame close to your back door. If the cold frame is nearby, you'll remember to go out and raise and lower the lid as needed to vent and heat the frame.

The ideal cold frame is 4 to 6 feet wide and about 3 feet from front to back. If it's much more than 3 feet long, it can be difficult to reach in to the back of the frame. Use a naturally rot-resistant

wood such as redwood, cedar, or cypress. You can also use ¼ to ½-inch plywood that has been coated with linseed oil.

Make the lid from an old window sash, clear acrylic plastic, or clear double-poly plastic nailed onto a wooden frame. The angle of the lid should be 35 to 45 degrees to capture the maximum amount of sunlight.

The frame can be set on the ground, but it's best to bury it a few inches for added insulation. To bury the frame, dig a hole 3 to 4 inches deep under the entire area where the frame is to be placed. Dig out an additional 2 inches on all sides. To control weeds within the frame, line the hole with a layer of black plastic or landscape fabric. Put the frame in place and add 1 to 2 inches of sand and/or gravel to help with water drainage and to keep the surface level. Replace the soil around the outside edges of the frame and tamp it firmly.

Make sure there is some kind of prop handy to keep the lid up on warm days. A stick and a piece of wood will both work, but they can easily be knocked over if the winds are heavy. A heavy object like a brick is a more reliable lid prop. The best way to

make sure the lid stays open is to position the cold frame against a wall, fence, shed, house, or other structure. Then screw a hook onto the back of the lid and an eye into the structure to hold the hook. To control the level of the opening, loop the hook into a chain; raise and lower the lid according to the weather conditions.

You can also purchase a ready-made cold frame from most garden-supply catalogs. Some models are equipped with automatic venting systems that are activated when the temperature inside the frame climbs above 70°F.

HOTBEDS

Another way to use a cold frame is to turn it into a hotbed. Hotbeds were traditionally made by packing a cold frame with fresh horse or cow manure. The manure heated up the frame quite effectively. Nowadays, fresh manure is not readily available to most of us, so electric heating cables placed in the bottom of the frame are used instead. The cables are covered with a 4 to 6-inch layer of a mixture of sand, soil, and peat moss.

Before making a hotbed with electric cables, do check with your utility company or a professional electrician to determine if you have the proper kind of cables and hookup equipment to prevent electrical shock and other safety hazards.

Seed can be directly sown into the hotbed soil and transplanted to other containers for growing on. Fall, late winter, and early spring crops of greens and cool-weather vegetables can be planted this way and grown to harvest in the frame.

4

ENCYCLOPEDIA OF
ORNAMENTAL ANNUALS
AND PERENNIALS

A

ABUTILON

Grow as an annual
Chinese bellflower, flowering maple

Sowing Directions: For planting outdoors, sow indoors 6–8 weeks before the last frost. For a houseplant, sow seed anytime of the year, but it will flower indoors only during the winter months. Lightly cover seeds with soil.

Abutilon 'Summer Sherbert'

Indoor Germination Temperature: 70–75°F

Days to Germination: 14–21

Growing On Temperature: 70–75°F

Garden Planting: Plant after all danger of frost has passed in full sun to part shade. Needs a consistently moist, rich, and well-drained soil. Space 2 feet apart.

Description: Fast-growing, vigorous plant with large 3–4-inch paperylike flowers in gorgeous shades of yellow, red, pink, white, and lavender. Height: 3 feet.

Recommended Cultivar: 'Summer Sherbert' displays 3–4-inch ivory, lemon, pink, rose, apricot, coral, or red flowers. Height: 2 feet.

ACHILLEA

Perennial
A. filipendula, A. Millefolium, A. Ptarmica, A. tomentosa
Yarrow, milfoil
Zones: 3–8

Sowing Directions: Sow indoors or outdoors 2 months before the last frost. Seeds need light to germinate.

Indoor Germination Temperature: 65–70°F

Days to Germination: 10–15

Growing On Temperature: 50°F

Garden Planting: Plant in full sun site. Prefers well-drained soil, but it will grow well in poor soils. Space 1–2 feet apart, depending on the variety.

Description: Hardy, drought-tolerant perennial with white, yellow, red, or multihued pastel flowers and fernlike foliage. Blooms in June and July; if cut back after initial flowering, it will usually bloom again in early fall. Excellent for borders and cut or dried flowers. Height: 2–4 feet.

Achillea 'The Pearl'

Recommended Cultivars:

'Coronation Gold': Extremely heat and drought tolerant. Height: 3 feet.

'Moonshine': Very bright yellow flowers and silvery gray foliage. Height: 2 feet.

'The Pearl': Masses of small, cotton-ball-like white flowers and dark green foliage. Height: 2½ feet.

'Summer Pastels': Blooms first year from seed. Pink, carmine, lilac, purple, bronze, silver, and white flowers. Height: 2 feet.

AGAPANTHUS AFRICANUS

Perennial
African lily, lily of the Nile
Zone: 9 (grow as houseplant in cold areas)

Sowing Directions: For planting outdoors, sow seeds indoors 2–3 months before the last frost. For a houseplant, sow seed anytime of the

NOTE: *African lilies grown from seed take 3–5 years to flower. The usual method of home garden growing is to purchase bulbs and plant them in the spring.*

year. Seeds need light to grow; cover very sparsely.

Indoor Germination Temperature: 70–75°F

Days to Germination: 30–35

Growing On Temperature: 65–70°F

Garden Planting: Plant in full sun or part-shade in well-drained soil. For best results, supply generous amounts of water and fertilizer during the growing season. In zones lower (colder) than Zone 8, plant in a container and place outdoors during the summer; bring it indoors to a cool (40–50°F) and dry location over the winter. Space 2 feet apart.

Description: Striking umbel-shaped clusters composed of many tubular flowers borne on tall flower stalks. Lower dark green, strappy leaves grow to 1 foot long. Height: 3–4 feet (in bloom).

Recommended Cultivars:

'Albus' has white flowers.

'Blue Danube' has light blue flowers.

'Bressingham Blue' has deep amethyst blue flowers.

Agapanthus africanus 'Blue Danube'

TIP:
Agapanthus grows very well in containers. Grow as houseplant in cold areas.

TIP:

Hollyhock often self-seeds; transplant the seedlings in the fall.

TIP:

Rust is a fungus disease that causes reddish brown spotting on the foliage. If rust becomes a problem, grow hollyhocks as annuals. Buy rust-resistant varieties.

AGERATUM HOUSTONIANUM

Annual
Ageratum, flossflower

Sowing Directions: For best results, sow indoors 6–8 weeks before the last frost. Sow outdoors after any danger of frost has passed. Seeds need light to germinate; do not cover with medium.

Indoor Germination Temperature: 75–82°F

Days to Germinate: 7–10

Growing On Temperature: 60–65°F

Garden Planting: Plant in full sun to light shade in rich well-drained soil. Keep well watered. Space 10–12 inches apart.

Description: Short, compact, bushy plant with clusters of blue, purple, or white flowers that bloom throughout the growing season. Best used as a container plant or massed as a border plant. Height: 4–30 inches.

Recommended Cultivars:

'Blue Danube' has lavender blue flowers. Height: 6 inches.

'Blue Horizon' has exceptionally intense blue flowers. Height: 30 inches.

ALCEA ROSEA

Biennial
Hollyhock
Zones: 3–7

Sowing Directions: Sow annuals and perennials indoors 8–10 weeks before the last frost date. Perennial varieties can be sown outdoors, anytime in spring or summer up to 2 months before the first fall frost, directly where they are to grow. Annuals will flower the first year; perennials will produce foliage the first season and flower the following year. Seeds need light to germinate.

Indoor Germination Temperature: 68–70°F

Days to Germination: 10–14

Growing On Temperature: 55–60°F

Garden Planting: Plant in full sun in rich, well-drained soil. Keep well watered and fertilize monthly during the growing season. Taller specimens should be staked if grown in a windy site. Space 15–20 inches apart.

NOTE: *When in full bloom, this glorious plant is a showstopper. All children should have a memory of hollyhocks growing in a summer garden.*

Description: Tall, erect plant with spires of old-fashioned-looking white, yellow, pink, red, lavender, or purple-black flowers that bloom from June through late August. Usually planted in the back of a border, but is especially charming if grown against a wall or fence. Susceptible to rust. Height: 3–9 feet.

Recommended Cultivars:

'Chaters Double' mix: An old classic with double flowers. Height: 6–8 feet.

'Nigra': Striking purple-black flowers. More resistant to rust than other varieties.

ALCHEMILLA MOLLIS

Perennial
Lady's mantle
Zones: 4–8

Sowing Directions: Sow indoors 2 months before the last frost. Use fresh seed; older seed should undergo a cold treatment. Best to sow outdoors, in fall or early spring, directly where plants are to grow. Established plants self-sow easily. Lightly cover seeds with soil.

Indoor Germination Temperature: 60–70°F

Days to Germination: 10–14

Growing On Temperature: 55–60°F

Garden Planting: Plant in full sun to light shade in moist, well-drained soil. In warmer zones, plant in part shade. Space 1½ feet apart

Description: Low-growing herb with 18-inch-tall stems that bear lime green flowers. The flowers, which bloom in late spring to early summer, make an unusual long-lasting cut flower. Use for the front of borders or plant in masses as a ground cover. It is known for the way early morning dew and rain droplets collect on the leaves like balls of silvery mercury. Height: 8–10 inches.

ALLIUM

Perennial (hardy bulb)
A. Christophii, A. giganteum, A. Moly, A. Schoenoprasum, A. senescens var. *glaucum*
Ornamental flowering onion
Zone: 4

Sowing Directions: Sow indoors 2 months before the last frost. Can be sown outdoors, in the fall, directly where they are to grow for spring germination.

Indoor Germination Temperature: 65–70°F

Days to Germination: 14–21

Growing On Temperature: 60°F

Garden Planting: Plant in full sun in loose, well-drained soil. Space 6–12 inches apart, depending on species.

Description: This plant has strappy, grasslike leaves with clusters of small star-shaped flowers that can form a ball on 8-inch to 5-foot stems borne above the foliage. Blooms appear in late spring to early summer. A very striking plant, especially *A. giganteum*, which can grow to 4 feet tall. Plant in masses for naturalizing a garden area. Plants started from seed should flower in 2–3 years. Height: 8 inches to 5 feet.

AMARANTHUS TRICOLOR; A. CAUDATUS

Annual
A. tricolor: Joseph's coat, fountain plant, tampala, summer poinsettia; *A. caudatus:* love-lies-bleeding, tassel flower

Sowing Directions: For best results, sow indoors 3–4 weeks before the last frost. Can be sown outdoors after the last frost.

Indoor Germination Temperature: 70–75°F

Days to Germination: 8–12

TIP:
Lady's mantle is commonly propagated by division. Divide in early spring before the plant starts to bloom.

TIP:
Flowering onion can be propagated by division. Dig up the plant, separate the bulbs, and replant. Divide spring-flowering bulbs in late summer, and summer-flowering bulbs in spring.

TIP:
Keep an eye on pearly everlasting in the garden border, as it may become invasive.

Joseph's coat amaranthus

Growing On Temperature: 65–68°F

Garden Planting: After all danger of frost, plant in full sun in a dryish, well-drained soil. Will not tolerate wet soils. Top-heavy plants benefit from staking just before flowering. Space 12–20 inches apart.

Description: Fast-growing plant with intense, brightly colored foliage. Its foliage is considered the most ornamental feature. Blooms midsummer to the first fall frost. Height: 2–4 feet.

A. tricolor: Foliage that displays a riotous mix of red, yellow, and green with insignificant flowers.

A. caudatus: Brilliant, glowing rosy red foliage with long tassellike red flowers.

Recommended Cultivar: 'Illumination': Intense red upper leaves with green and chocolate brown lower leaves. Height: 3 feet.

AMARYLLIS. SEE *HIPPEASTRUM*

ANAPHALIS MARGARITACEA

Perennial
Pearly everlasting
Zones: 3–8

Sowing Directions: Sow indoors 8–12 weeks before the last frost. Can be sown outdoors in fall or early spring.

Indoor Germination Temperature: 55–65°F

Days to Germination: 10–14

Growing On Temperature: 55–60°F

Garden Planting: Plant in full sun to part shade in well-drained, dryish soil. Space 12–14 inches apart.

Description: This hardy, drought-tolerant perennial displays gray-green foliage and small fluffy white flowers with yellow centers that bloom in July and August. Excellent for dried or fresh-cut flower arrangements. Height: 2–3 feet.

ANCHUSA

Annual; perennial
A. capensis: (annual) summer forget-me-not; *A. azurea:* (perennial) Italian alkanet, bugloss
Zones: 3–8

Sowing Directions: Sow annuals indoors 6–8 weeks before the last frost; can be sown directly outdoors after the last frost. Sow perennials anytime in spring or summer up to 2–2½ months before the first fall frost. Perennials must be sown indoors in early spring to produce flowers the first year.

Indoor Germination Temperature: 70–75°F

Days to Germination: 15–20

Growing On Temperature: 55–60°F

Garden Planting: Plant in full sun to part shade in any well-drained soil. Plant perennials in poor soil and do not fertilize. Space annuals 9–12 inches apart; perennials, 2 feet apart.

Description:

A. capensis is low growing, with a sprawling habit. A profuse display of lovely, tiny azure flowers appears from June or July to the first frost. Height: 7–12 inches.

A. azurea has masses of tiny blue flowers that bloom in June and July. Stake specimens that grow taller than 4 feet. Can be invasive, but massing produces an informal natural design. Height: 3–5 feet.

Recommended Cultivar: 'Dropmore' has bright blue flowers and grows 2½–3 feet in height.

ANEMONE

Perennial (hardy bulb)
A. blanda: Greek anemone, Greek windflower, Grecian windflower; *A. coronaria:* poppy anemone, lily-of-the-field (Zone 8); *A.* × *hybrid*a: Japanese anemone; *A. Pulsatilla:* pasqueflower
Zone: 5–10

Sowing Directions: Sow indoors 2 months before the last frost. *A.* × *hybrida* and *A. Pulsatilla* can be sown outdoors in the fall. Seeds need light to germinate; cover sparsely with medium.

Indoor Germination Temperature: 65–70°F

Days to Germination: 21–28

Growing On Temperature: 60–65°F

Garden Planting: Most species prefer full sun; *A.* × *hybrida* and *A. blanda* can take part shade. All need a well-drained soil amended with lots of organic matter. Most prefer cool growing conditions; in warm climates, make sure plants receive adequate moisture. Space *A. blanda* and *A. Pulsatilla* 8–12 inches apart; and *A.* × *hybrida* 1½ feet apart.

Anemone blanda

Description: Very cheering with bright, multicolored poppylike flowers.

A. blanda: Low growing with sky blue, pink, red, or white flowers. Height: 3–6 inches.

A. coronaria: Single or double flowers of red, pink, white, or blue that bloom mid to late spring into early summer. Height: 18 inches.

TIP:
Cut back perennials after flowering to encourage more flowers in the fall. For a cottage garden look, interplant with white daisies.

TIP:

A. × hybrida is one of the more beautiful and showy fall-blooming perennials. Plant in the border for a summerlike burst of color in the fall.

TIP:

For fuller branching and thus more flowers, it is important to pinch back young plants. Deadheading will also increase bloom.

TIP:

Most snapdragons do not bloom all summer. Therefore, plant them in a mixed border rather than massed by themselves. Single-color varieties are good in mixed border plantings.

A. × hybrida: Graceful, swaying white flowers borne above the foliage. Blooms late summer through to frost. Height: 2–3 feet.

A. Pulsatilla: Erect flower stems that bear blue to reddish purple flowers. Height: 10–12 inches.

Recommended Cultivars:

A.× hyrida 'Alba': Single white flowers. Height: 2 feet.

'September Sprite': Single light pink flowers. Height: 30 inches.

'Whirlwind': Semidouble white flowers. Height: 3 feet.

ANTIRRHINUM MAJUS

Grow as an annual
Snapdragon

Sowing Directions: For optimum germination, chill seeds in a refrigerator for 3–4 days before sowing. Sow indoors 8–12 weeks before the last frost. Mature plants sometimes self-sow, but hybrids will not be true to the mother plants. Seeds need light to germinate; do not cover with medium.

Indoor Germination Temperature: 65–70°F

Days to Germination: 7–14 days

Growing On Temperature: 50–55°F

Garden Planting: Plant in full sun in well-drained sandy soil. If the soil is not naturally fertile, apply a slow-release fertilizer early in the growing season. It prefers cool weather; in warm climates, plant heat-resistant varieties. Space 6–8 inches apart; space taller varieties farther apart.

Description: Uniquely shaped flowers in a range of yellow, red, maroon, pink, and white that bloom in early summer. Use in borders, beds, and edgings and for cut flowers. Very susceptible to rust; plant rust-resistant varieties and/or regularly apply a fungicide. Height: 6 inches (dwarfs) to 3 feet.

Antirrhinum majus 'Burpee's Double Supreme' hybrid mix

Recommended Cultivars:

'Burpee's Double Supreme' hybrid mix: Double flowers in a mix of colors. Heat resistant. Height: 2½ feet.

'Burpee's Topper' hybrid mix: Vigorous and heat resistant. Height: 3 feet.

'Cinderella' mix: Semidwarf; early bloomer. Height: 20 inches.

AQUILEGIA

Perennial
A. caerulea (Zones 3–8); *A. flabellata* (Zones 5–8); *A. × hybrida* (Zones 4–8)
Columbine

Sowing Directions: For indoor sowing, chill seeds in a refrigerator (40°F) for 2–3 weeks. Sow treated seeds 8 weeks before the last frost. For outdoor sowing, sow anytime in early spring through summer up to 2 months before the first fall frost. Plants grown from seeds sown in early spring may not flower the first year. Once established, plants will readily self-sow and, if not a hybrid, will usually grow true to type. Seeds need light to germinate; do not cover.

Indoor Germination Temperature: 70–75°F

Days to Germination: 21–28

Growing On Temperature: 50–55°F

Garden Planting: Plant in full sun to part shade in moist, fertile, well-drained soil. In hot and dry climates, plant in part shade. Space 12–15 inches apart.

Description: One of the most graceful and distinctive garden flowers with light, airy foliage and spurred multicolored sepals and petals. Flowers are red, mauve, white, blue, pink, yellow, or purple and bloom in late spring to early summer. A. flabellata is a compact species, and A. × hybrida includes some of the most popular and showy varieties. Height: A. caerulea, 18–30 inches; A. flabellata, 18 inches; A. × hybrida, 1–3 feet.

Recommended Cultivars:

'Harlequin' mix: Blooms earlier and more profusely than other cultivars. The tallest mix in a range of colors. Height: 3 feet.

'McKana Giant' mix: A rainbow of very large blooms with long spurs. Height: 2½ feet.

'Songbird' hybrid mix: Large, multicolored flowers (3–3½ inches) borne on compact plants. Height: 18–24 inches.

ARCTOTIS STOECHADIFOLIA

Annual
African daisy, arctotis

Sowing Directions: For best results, sow indoors 6–8 weeks before the last frost. Can be sown outdoors 3–4 weeks before the last frost.

Indoor Germination Temperature: 60–70°F

Days to Germination: 21–25

Growing On Temperature: 60–65°F

Garden Planting: Plant in full sun in well-drained to dry, sandy soil. Space 10–12 inches apart.

Description: A drought-tolerant plant with brilliant creamy white, red, yellow, pink, orange, or brownish red flowers. Height: 10–24 inches.

ARMERIA MARITIMA

Perennial
Sea pink, sea thrift, common thrift
Zones: 3–7

Sowing Directions: Soak seeds in warm water overnight (6–8 hours) before sowing. For best results, sow indoors 2 months before the last frost. Can be sown outdoors anytime in spring or summer up to 2 months before the first fall frost.

Indoor Germination Temperature: 68–70°F

Days to Germination: 14–21

Growing On Temperature: 50–55°F

TIP:
Plant with tulips and other early spring–flowering bulbs. Columbine blooms after those bulbs have finished flowering, and its foliage helps hide the bulbs' spent foliage.

TIP:

Goatsbeard makes an excellent herbaceous specimen plant for perennial borders or island beds.

Armeria maritima

Garden Planting: Plant in full sun in well-drained to sandy soil. Thrives in seaside conditions, thus its common name. Will rot if planted in fertile, moist soils. Space 8–10 inches apart.

Description: Blooms in May and June and is most often used in rock gardens and in the front of borders. Height: 6–12 inches.

ARUNCUS

Perennial
Zones: 4–7
A. dioicus
Goatsbeard

Sowing Directions: Seeds do best if given a cold treatment at 40°F for 2–3 months. Sow seeds 2 months before the last frost. Seeds can also be sown in early summer; in the fall or the following spring, transplant young plants to where they are to grow.

Indoor Germination Temperature: 60–65°F

Days to Germination: 14–21

Growing On Temperature: 55–60°F

Garden Planting: Prefers part shade but will withstand full sun in cool climates if given steady moisture. Plant in moist, well-drained soil and provide generous amounts of organic matter. Space 4 feet from all other plants.

Description: Large, shrublike plant with cream-colored flower spikes in early summer. Height: 4–6 feet.

ASCLEPIAS TUBEROSA

Perennial
Butterfly weed, pleurisy root
Zones: 4–9

Sowing Directions: Sow indoors 2 months before the last frost. Sow outdoors, in early spring or fall, directly where they are to grow. If germination proves disappointing, chill the seeds at 36–40°F for 3 weeks before sowing.

Indoor Germination Temperature: 65–70°F

Days to Germination: 21–28

Growing On Temperature: 60–65°F

Garden Planting: Plant in full sun in

Asclepias tuberosa

dryish, well-drained soil. Will rot in wet soils. Does not like to be transplanted. Space 10–12 inches apart.

Description: A tough, drought-resistant native wildflower with bright orange flowers that attract butterflies. Blooms all summer to early fall. Good for border plantings. Height: 2–3 feet

Recommended Cultivar: 'Gay Butterflies' has gorgeous red and yellow flowers.

ASTER

Perennial
Aster, Michaelmas daisy
Zones: 3–8

Sowing Directions: For best results, sow indoors 2 months before the last frost. Sow outdoors, from late spring up to 2 months before the first fall frost, directly where they are to grow.

Indoor Germination Temperature: 70–75°F

Days to Germination: 15–20

Growing On Temperature: 60–62°F

Garden Planting: Most need full sun and well-drained soil. Plants need good air circulation around them to avoid powdery mildew problems. Space 3 feet apart for tall varieties (higher than 3 feet); 1–1½ feet apart for short varieties.

Description: Displays a profusion of daisylike flowers in purple, red, pink, blue, and white. Most species bloom in late summer or early fall through to frost. *A. alpinus* blooms in May and June. Height: 2–7 feet; *A. alpinus*, 1 foot.

Recommmended Cultivars:

'Harrington's Pink': Abundant blooms of pink flowers. Height: 3–4 feet.

'Wonder of Staffa': Compact with lovely blue flowers. Height: 1½–2 feet.

ASTILBE × ARENDSII

Perennial
Astilbe, false spirea
Zones: (3) 4–8

Sowing Directions: Sow seeds indoors 2 months before the last frost. Can be sown outdoors, in early spring, where they are to grow. Be patient, these seeds are slow to germinate, grow on, and develop.

Indoor Germination Temperature: 60–70°F

Days to Germination: 21–30

Growing On Temperature: 60–65°F

Astilbe × Arendsii

TIP:

In addition to chrysanthemums and *Sedum* 'Autumn Joy', asters are a classic source of fall color for flower beds and container plantings.

TIP:

Pinch back fall-blooming asters in June for fuller plants. Stake tall-growing species in late summer.

TIP:
Low-growing
Astilbe chinensis
'Pumila' makes a
lovely ground
cover for moist,
shady sites.

TIP:
Shear back
plants after flow-
ering each year
so they retain a
rounded form.

Garden Planting: Plant in part shade in rich, well-drained, moist soil. Keep it well watered during the hot summer months. Will tolerate full sun if kept well watered. Space 1½–2 feet apart.

Description: Fernlike foliage is green to bronzy green; feathery flower plumes are red, pink, white, or lavender-pink and bloom in mid-summer through early fall. One of the best flowering perennials for shady sites. Height: 1½–3 feet.

Recommended Cultivars:

'Deutschland': Early-blooming with creamy white flowers. Height: 2 feet.

'Fanal': Deep red flowers. Height: 2½ feet.

'Sprite': Bright green foliage with light pink flowers. Height: 18–20 inches.

Aubrieta deltoidea

Aubrieta deltoidea

Perennial
False rock cress, purple rock cress
Zones: 4–7

Sowing Directions: Sow seeds indoors 2 months before the last frost. Can be sown outdoors in early spring to summer up to 2 months before the first fall frost. Plants started from seed will bloom the second year. Seeds need light to germinate; lightly press onto surface and do not cover with medium.

Indoor Germination Temperature: 65–70°F

Days to Germination: 14–21

Growing On Temperature: 50–55°F

Garden Planting: Plant in full sun to part shade in rich, well-drained soil. Space 8–10 inches apart.

Description: A low-growing, spreading plant with dazzling pink, light purple, or red flowers that bloom in early to late spring. Most often planted in rock gardens, border edges, and containers and on top of low garden walls and terrace edges. Height: 6–8 inches.

B

BABY'S BREATH. SEE *GYPSOPHILA*

BALLOON FLOWER. SEE *PLATYCODON GRANDIFLORUS*

Baptisia australis

Perennial
False indigo, blue indigo
Zones: 3–8

Sowing Directions: For sowing indoors, first chill seeds in the refrigerator (40°F) for 6 weeks. Then scratch the rough seed coat with a nail file or nick with a sharp knife. Sow 2 months before the last frost. Sow outdoors, in the fall or early spring, directly where plants are to grow.

Indoor Germination Temperature: 70–75°F

Days to Germination: 5–10

Growing On Temperature: 60°F

Garden Planting: Plant in full sun to part shade in any soil. Cut plant back by one-third to one-half after it flowers to retain a fuller habit. Stake taller specimens. Space 1½–2 feet apart.

Description: A tall-growing, shrub-like plant with indigo blue flowers in June and early summer. Foliage is an attractive blue-green color and is ornamental on its own. Seedpods form and hang on plant after flowering. Very easy to grow; may become invasive since it readily self-sows. Height: 4–6 feet.

BEE BALM. SEE *MONARDA DIDYMA*

BEGONIA

Annual
Begonia; wax begonia, bedding begonia; tuberous begonia, rex begonia

Sowing Directions: For planting outdoors, sow seed 3–3½ months before the last frost. This very fine seed is difficult to sow evenly and is often sold pelleted. For a house-plant, sow any time of year. Plants grown from seed will bloom in 3–4 months. Seeds need light to germinate; lightly press onto the well-moistened potting mix and do not cover with medium. Cover the container so seeds do not dry out; mist frequently.

Indoor Germination Temperature: 78–80°F

Days to Germination: Wax, 14–21; tuberous, 15–30

Growing On Temperature: 60°F

Garden Planting: Plant wax begonias in full sun to dense shade in almost any kind of soil, although they prefer rich, well-drained soil. Fertilize monthly. If growing in full sun in warm climates, provide adequate water. Space 10–12 inches apart. Plant tuberous and rex begonias in part shade in a rich, moist, well-drained soil. Plant 12–18 inches apart.

Description: Flower colors include pink, red, white, yellow, and orange. There are three principal types of begonia.

B. × *semperflorens-cultorum:* Wax, or fibrous-rooted begonias, are known as a versatile, reliable bedding plant. The waxy leaves are dark green to bronzy green. The white, pink, or red flowers bloom all summer and into the fall. Height: 6–12 inches.

B. × *tuberhybrida:* Tuberous begonias develop tuberous roots and are larger and more tropical looking than wax begonias. Dazzling 4–6-inch single and

TIP: Handle wax begonias very carefully when transplanting. Their brittle stems and leaves break easily.

TIP: Tuberous and rex begonias are often grown in containers indoors in the colder months and outside during the summer. If planting tuberous begonias directly in the ground, dig up the tubers before the first fall frost. Store in a cool, dry place and replant in the spring.

double flowers in white, yellow, pink, orange, or apricot bloom all summer. Grow as an annual or houseplant. Height: 1–3 feet.

Rex begonias, or rhizomatous begonias, are grown primarily as houseplants and for their ornamental foliage. Most begonias are hardy only in Zone 9 or 10. The exception is hardy begonia *(B. grandis),* which is hardy to Zone 7. Rex and hardy begonias are most easily propagated through cuttings or division. Hardy begonias readily self-sow by bulbils in the spring.

TIP:

Mulch plants over the winter since blackberry lily is susceptible to frost heave.

Begonia 'Wings' hybrid mix

Recommended Cultivars:

'Memory' hybrid mix (tuberous begonia): Double (6½-inch) flower in a mix of colors.

'Nonstop' hybrid mix (tuberous begonia): Medium (3–4-inch) flowers. Seed often sold pelleted for easier handling.

'Wings' hybrid mix (wax begonia): Large (3-inch) flowers bloom in 3 months from seed in a variety of colors.

BELAMCANDA CHINENSIS

Perennial
Blackberry lily, leopard flower
Zones: 5–10

Sowing Directions: Sow seeds indoors about 2 months before the last frost. Can be sown outdoors in early spring up to 2 months before the first fall frost. Plants grown from seed bloom in the second year. Established plants readily self-sow.

Indoor Germination Temperature: 70–85°F

Days to Germination: 15–20

Growing On Temperature: 60–65°F

Garden Planting: Plant in full sun to light shade in rich, well-drained soil. Likes consistent moisture, especially during the hot summer months. Space 8–10 inches apart.

Description: Orange-colored, lily-like flowers with flecks of red spots bloom in late summer. After flowering, clusters of small black seedpods form that look like blackberries. Foliage resembles that of the iris. Height: 2–4 feet (in bloom).

BELLFLOWER. SEE CAMPANULA

BELLIS PERENNIS

Biennial
English daisy
Zones: 4–8

Sowing Directions: Sow seeds indoors 6–8 weeks before the last frost. For outdoor sowing, sow into cold frames in late summer and transplant to the garden the following spring. Or sow outdoors, in early spring, directly where they are to grow.

Indoor Germination Temperature: 70–75°F

Days to Germination: 7–14

Growing On Temperature: 50–55°F

Garden Planting: Plant in full sun to part shade in rich, moist, well-drained soil. It prefers cool growing conditions; mulch to keep soil cool and moist. Grow as an annual in Zones 9–11. Will not tolerate hot weather without a cool winter period. Space 8–10 inches apart.

Description: White, red, or pink daisy flowers with yellow centers bloom in May and June. Can be invasive. Height: 6–8 inches (when in bloom).

BELLS OF IRELAND. SEE
MOLUCCELLA LAEVIS

BERGENIA CORDIFOLIA

Perennial
Bergenia, heart-leaf bergenia
Zones: 3–8

Sowing Directions: Sow indoors 2–3 months before the last frost. If germination is poor, chill the seeds in a refrigerator (32–41°F) for 2–3 weeks and sow again. Can be sown directly outdoors in early spring or late fall.

Indoor Germination Temperature: 55–60°F

Days to Germination: 15–20

Growing On Temperature: 55°F

Garden Planting: Plant in part shade in rich, moist, well-drained soil. It will tolerate sun if the soil is kept moist. Space 1–1½ feet apart.

Description: Low-growing plant with heart-shaped, thick, glossy evergreen leaves that have a reddish purple tint in the winter. Pink, red, or white flowers bloom in April and May. Most often used in part shade as a ground cover or in the front of the border. Watch for slugs, which can make the foliage look quite ratty. Height: 1 foot.

Recommended Cultivar: 'Redstart' has red flowers and is a good cultivar to start from seed.

BLACKBERRY LILY. SEE
BELAMCANDA CHINENSIS

BLANKETFLOWER. SEE
GAILLARDIA

BLAZING STAR. SEE
LIATRIS SPICATA

BLEEDING HEART. SEE
DICENTRIS SPECTABILIS

BLUEBELLS, VIRGINIA.
SEE MERTENSIA VIRGINICA

BLUE LACE FLOWER. SEE *TRACHYMENE COERULEA*

BOUGAINVILLEA

Perennial (tender), grow as an annual
Bougainvillea, paper flower
Zone: 10

Sowing Directions: Seed can be sown indoors anytime of the year.

Indoor Germination Temperature: 70–75°F

Days to Germination: 30–35

Growing On Temperature: 70–80°F

Garden Planting: Most often planted in containers in rich, well-drained, moist soil and given trellis support. Can be grown in full sun outdoors or in a southern exposure window indoors or in a greenhouse. Fertilize monthly and water generously during active growth. Prune back to keep the plant fuller looking.

Description: Luxuriant, vigorous-growing, tropical vine with ravishing magenta, yellow, red, purple, white flowers. Height: 2–20 feet.

BRACHYCOME IBERIDIFOLIA

Annual
Swan River daisy

Sowing Directions: For best results, sow seeds indoors 6–8 weeks before the last frost. In climates with a longer growing season, can be sown, after all danger of frost has passed, directly where they are to grow. Successive direct sowings of seed

every 3 weeks, beginning in early spring, allow for continuous bloom from summer to fall.

Indoor Germination Temperature: 70°F

Days to Germination: 10–18

Growing On Temperature: 55–60°F

Brachycome iberidifolia, mixed colors

Garden Planting: Plant in full sun in rich, well-drained soil. Can tolerate some shade. Deadhead flowers to prolong bloom time. Space 8–10 inches apart.

Description: Displays a profusion of dainty, daisy-like white to lavender flowers with a light, fresh scent. Excellent bedding or container plant. Height: 6–10 inches.

BRASSICA OLERACEA ACEPHALA GROUP

Annual
Flowering kale, flowering cabbage

Sowing Directions: For fall and winter bloom, sow seeds indoors in a moist, soilless mix and place in a refrigerator for 2–3 days. After the cold treatment, the potting medium should be allowed to warm to about 65°F. Can sow outdoors in a shaded area in midsummer (June through July). Seeds need light to germinate; do not cover with medium.

Indoor Germination Temperature: 65–70°F

Days to Germination: 7–14

Growing On Temperature: 55–60°F

Garden Planting: Plant in full sun in moist, well-drained soil. Space about 12 inches apart.

Description: Large ornamental rosettes of green, reddish purple, and purple cabbage leaves. Foliage becomes more colorful after the first frost. Widely used as a fall and winter outdoor container plant.

BRIZA MAXIMA

Annual
Quaking grass

Sowing Directions: Can be sown indoors, if soil temperature is held to 55°F. Sow 3–4 seeds to a small container and do not transplant to an intermediate pot before garden planting. It is easier to sow outdoors, in early spring, directly where the plants are to grow.

Indoor Germination Temperature: 55°F

Days to Germination: 10–14

Growing On Temperature: 58–60°F

Garden Planting: Plant in full sun in any well-drained to sandy soil. Space at least 12 inches apart.

Description: An ornamental grass

with seedpods that resemble elongated hearts and gently nod in the breeze. A short-lived plant (if planted in the garden in May, it will be dying by the end of July). Commonly used for fresh and dried flower arrangements. Height: 2 feet.

BROWALLIA SPECIOSA

Browallia speciosa

Annual
Browallia

Sowing Directions: Sow indoors 8–10 weeks before the last frost (plants grown from seed take a long time to bloom). Pinch the tops of seedlings when they reach 5 inches to encourage a bushier habit. Seeds need light to germinate; scatter and press lightly onto surface and do not cover with medium.

Indoor Germination Temperature: 75°F

Days to Germination: 7–15

Growing On Temperature: 60–65°F

Garden Planting: Plant in full to part shade in rich, well-drained soil. In warmer climates, it will do best in part shade. Provide plenty of water if growing in full sun and/or in containers. Foliage benefits from an occasional misting throughout the growing season. Can be grown indoors as a houseplant if given a sunny southern window. Space 10–12 inches apart.

Description: Charming star-shaped blue to light purple and white flowers bloom all summer until the first frost. It has a very neat, well-shaped appearance. Often used as a container or hanging basket plant. Height: 10–18 inches.

C

CABBAGE, FLOWERING. SEE *BRASSICA OLERACEA ACEPHALA GROUP*

CALENDULA OFFCINALIS

Annual
Pot marigold

Sowing Directions: In frost zones, sow seeds indoors 6–8 weeks before the last frost. Or after the last hard frost and 2–3 weeks before the last frost, sow directly where they are to grow. In frost-free areas, sow in early fall and winter since plant will not thrive in intense summer heat. Seeds need darkness to germinate; cover completely with medium.

Indoor Germination Temperature: 70°F

Days to Germination: 7–14

Growing On Temperature: 55–60°F

Garden Planting: Plant in full sun in well-drained, fertile soil that is kept well watered. It prefers cool growing conditions. Space 12–16 inches apart.

Description: A colorful, bushy plant with marigold- or chrysanthemum-like orange, yellow, or white flowers that bloom from early summer to the first fall frost. In warm climates, flowers will fade out by the end of summer, but can be planted for spring and fall display. Height: 8–24 inches.

Calendula 'Dwarf Gem' in orange and gold.

Recommended Cultivar: 'Dwarf Gem' mix is a compact plant with 3-inch double flowers in lemon yellow, apricot, gold, and orange. Terrific for edging and containers.

CALIFORNIA POPPY. SEE *ESCHSCHOLZIA CALIFORNICA*

CALLISTEPHUS CHINENSIS

Annual
Annual aster, China aster

Sowing Directions: Sow indoors 6–8 weeks before the last frost. Or sow outdoors, after the last frost, where they are to grow. Successive sowings at 2–3-week intervals will allow a continuous harvest of blooms for indoor cut flowers.

Germination Temperature: 70°F

Days to Germination: 8–14

Growing On Temperature: 60–62°F

Garden Planting: Plant in full sun in fertile, well-drained soil. Keep well watered throughout the growing season. Space 8–16 inches apart.

Description: Grown primarily for cut flowers, which can be either single or double and come in a wide spectrum of colors: purple, pink, red, lavender, and white. Susceptible to wilt. Height: 12–30 inches.

Recommended Cultivars:

'Burpee's Red Mound': Compact cultivar that produces a profusion of deep scarlet, 3-inch double flowers. Height: 14 inches.

'Fireworks' mix: Bushy plant with quilled flowers in rose, scarlet, pink, blue, and white. Height: 24 inches.

'Totem Pole' mix: Displays a gorgeous blend of blue, cerise, rose-pink, and white large (5-inch) double flowers. Height: 2 feet.

CAMPANULA

Biennial; perennial; can be grown as an annual
C. carpatica: (perennial) Carpathian harebell, Carpathian bellflower, Tussock bellflower (Zone 3); *C. glomerata:* (perennial) Dane's blood bellflower, clus-

tered bellflower (Zone 3); *C. Medium* var. *calycanthema:* (biennial) Canterbury bells; *C. persicifolia:* (perennial) peach-leaf bellflower, willow bellflower, paper bellflower (Zones 5–8); *C. Portenschlagiana:* (perennial) dalmatian bellflower (Zones 5–8)

Campanula glomerata

Sowing Directions: For all species, sow seed indoors 8 weeks before the last frost. Can also be sown, in late spring or early summer, directly where they are to grow for bloom in the following season. Seeds need light to germinate, press lightly onto the surface and do not cover with medium.

Germination Temperature: 70°F

Days to Germination: 14–21

Growing On Temperature: 55–60°F

Garden Planting: Plant in full sun to part shade in well-drained soil. Keep well watered and apply mulch to

TIP:
To avoid wilt problems, choose wilt-resistant varieties and do not plant annual aster in the same spot each year.

keep the roots cool. Space 6–12 inches apart, depending on cultivar.

Description: The lovely bell- or cup-shaped white, blue, or blue-violet flowers bloom from late spring to early summer.

C. carpatica: Mounded habit with dainty cup-shaped blue or white flowers that bloom from early summer to early fall. Height: 12 inches.

C. glomerata: Erect habit with terminal clusters of white or blue flowers. Height: 1–3 feet.

C. Medium: Erect habit and raceme flowers. Height: 2–3 feet.

C. persicifolia: Erect habit and terminal raceme white, blue, or pink flowers. Height: 2–3 feet.

C. Portenschlagiana: Low growing with funnel-shaped lilac-blue flowers. Height: 4–8 inches.

Recommended Cultivars (*C. carpatica*):

'Blue Clips' has medium blue cup-shaped flowers. Terrific for a rock garden, edgings, or the front of a border. Height: 10 inches.

'White Clips' is a pure white version of 'Blue Clips'.

CANDYTUFT. SEE *IBERIS SEMPERVIRENS*

CANNA × GENERALIS

Perennial (tender), grow as an annual
Canna

Sowing Directions: Sow seeds indoors 6–8 weeks before the last frost. Seeds will germinate better if they are scarified and soaked in warm water overnight before sowing.

Germination Temperature: 70–75°F

Days to Germination: 8–12

Growing On Temperature: 65°F

Garden Planting: Although, easy to grow from seed, they are usually planted as rhizomes. Plant in full sun in rich, well-drained soil. Keep well watered and fertilize during the growing season. In frost areas, dig up the rhizomes after the first frost and store in a cool, dry location for planting the following spring. Storage temperature should not get below 40°F. Space 12–24 inches apart.

Description: Tropical-looking plant with lush clusters of large, exotic red, rose, yellow, orange, gold, or white flowers on strong, erect stems. Height: 8 feet (large); 3 feet (dwarf).

Canna × generalis 'Tropical Rose'

Recommended Cultivar: 'Tropical Rose'—dwarf variety that is easy to start from seed. Soft rose flowers bloom 90 days after seed is sown. Excellent for large container planting. Height: 3 feet.

CANTERBURY BELLS. SEE *CAMPANULA*

CAPSICUM ANNUUM. SEE ORNAMENTAL PEPPER (IN CHAPTER 5)

CARDINAL FLOWER. SEE *LOBELIA CARDINALIS*

CARNATION. SEE *DIANTHUS*

CATANANCHE CAERULEA

Perennial
Cupid's dart
Zones: (3) 4–9

Sowing Directions: For bloom the first season, sow seed indoors 8 weeks before the last frost. Can be sown directly outdoors after the last frost.

Germination Temperature: 65–70°F

Days to Germination: 21–25

Growing On Temperature: 60°F

Garden Planting: Plant in full sun in well-drained soil. It will not tolerate wet soils and is drought tolerant. Space 10 inches apart.

Description: Plants forms a rosette of leaves with tall flower stalks that bear 2-inch lavender-fringed flowers that bloom in June and July. Height: 24 inches (in bloom).

CATNIP. SEE *NEPETA* (IN CHAPTER 6)

CATHARANTHUS ROSEUS (VINCA ROSEA)

Grow as an annual
Periwinkle, vinca

Sowing Directions: Seeds must be sown indoors. Sow 10–12 weeks before the last frost. Seeds need darkness to germinate; cover completely with medium.

Germination Temperature: 75–80°F

Days to Germination: 14–20

Growing On Temperature: 65–70°F

Garden Planting: Plant in full sun to part shade in any well-drained soil. It will tolerate hot, dry conditions but thrives if kept well watered during the hot summer months. Space 10–12 inches apart.

Description: Low-growing plant with glossy dark green leaves and brightly colored pink, red, white, or lilac flowers that bloom from late spring to frost. Grow as a perennial in Zones 9–11. Height: 4–12 inches.

Vinca rosea 'Arizona' mix

TIP:
Plant Cupid's dart in masses for the most ornamental and natural look.

TIP:
Because cornflower blooms are not long lasting, it is a good idea to do successive sowings. Sow at 2-week intervals until early summer for continuous bloom throughout the summer.

Recommended Cultivar: 'Arizona' mix includes pink, white, red, and lilac flowers. This Burpee exclusive is the first red-flowering variety. Height: 12 inches.

CELOSIA

Annual
C. cristata: celosia, crested cockscomb; *C. cristata* Plumosa Group: plumed celosia

Sowing Directions: Sow seed indoors 4–5 weeks before the last frost. Can be sown outdoors, after the last frost, where they are to grow. Mature plants self-sow. Seeds need light to germinate; barely cover with medium and mist daily.

Indoor Germination Temperature: 70–75°F

Days to Germination: 8–10

Growing On Temperature: 65–70°F

Garden Planting: Plant in full sun in any well-drained, sandy soil. Space 8–18 inches apart, depending on variety.

Description: This drought- and heat-tolerant plant displays intensely colored orange, red, yellow, and purple flowers that bloom throughout the growing season. Flowers are long

Celosia 'Floradale' series

lasting when cut and can be dried. Height: 6 inches–4 feet.

C. cristata has large, rounded, brain-shaped flowers with a soft, velvety texture.

C. cristata Plumosa Group has soft, wheatlike, feathery plumes.

Recommended Cultivars:

'Floradale' series is made up of crested varieties.

'Wine Sparkler' is a plumed variety with jewellike, richly red flowers. Height: 30 inches.

Centaurea

CENTAUREA CYANUS

Annual
Bachelor buttons, cornflower

Sowing Directions: Sow seed indoors 4 weeks before the last frost. For best results, sow outdoors, after the last frost, directly where they are to grow. Mature plants will often self-sow. Seeds need darkness to germinate; cover completely with medium.

Germination Temperature: 65–70°F

Days to Germination: 7–14

Growing On Temperature: 50–55°F

Garden Planting: Plant in full sun in well-drained soil. Space 6–12 inches apart.

Description: This drought- and heat-tolerant plant is best-known for its natural habit and old-fashioned blue flowers. It blooms from May to July and is also available with pink, mauve, and maroon flowers. Height: 12–30 inches.

CENTAUREA MONTANA

Perennial
Perennial cornflower, perennial
bachelor buttons, mountain bluet
Zones: 3–8

Sowing Directions: Sow seeds into containers in early to midsummer and transplant to final growing location in early fall or the following spring.

Germination Temperature: 70–75°F

Days to Germination: 7–14

Growing On Temperature: 55–60°F

Garden Planting: Plant in full sun to part shade in any well-drained soil. Space 12–15 inches apart.

Description: This erect plant has 2-inch-wide fringed blue flowers that bloom in May and June. Cut back plants after flowering for a second bloom period in early fall. Can become invasive if the established plants are not divided every 2 years. Height: 12–24 inches.

CHAMOMILE. SEE *CHAMAEMELUM NOBILE* (*ANTHEMIS NOBILE*) (IN CHAPTER 6)

Chrysanthemum

CHRYSANTHEMUM COCCINEUM

Perennial
Painted daisy, pyrethrum
Zones: 2–9

Sowing Directions: Sow indoors 8 weeks before the last frost. Or sow outdoors, after the last frost, directly where they are to grow.

> **NOTE:** *If seeds are sown indoors in January, they will flower in July of the same year. If sown later in the spring, they will bloom the following year.*

Germination Temperature: 60–70°F

Days to Germination: 14–21

Growing On Temperature: 55–60°F

Garden Planting: Plant in full sun in any well-drained soil. Keep the soil well watered, especially in warmer climates. Cut back after flowering for sporadic late summer and fall blooms. Space 10–12 inches apart.

Chrysanthemum coccineum

Description: Brightly colored red, pink, orange, and white daisylike flowers that bloom in May and June. Ferny foliage. Flowers are used to make the natural insecticide pyrethrum. Height: 2 feet.

CHRYSANTHEMUM × MORIFOLIUM

Perennial
Chrysanthemum, hardy chrysanthemum
Zones: 5–9

Sowing Directions: Sow seeds 8 weeks before the last frost. Seed sown at this time produces plants that bloom the first year.

Germination Temperature: 60–70°F

Days to Germination: 5–10

Growing On Temperature: 55–60°F

Garden Planting: After the last frost, plant in full sun in well-drained soil. Keep the soil well watered and fertilized. Pinch back flower buds until mid-July for bushier growth. Space 12–24 inches apart.

Description: This plant has a cushion-type habit and produces its classic fall flower in a variety of shapes, including pompom, cactus, spider, and daisy, in red, bronze, pink, yellow, white, lavender, and orange. Flowers bloom from late summer to first frost. Height: 8 inches–4 feet.

Recommended Cultivar: 'Autumn Glory' hybrid mix has double and semidouble yellow, orange, bronze, and red flowers. It will bloom first year from seed if started early. Height: 16 inches.

CHRYSANTHEMUM PARTHENIUM

Perennial
Feverfew, matricaria
Zones: 4–11

Sowing Directions: Sow seed indoors 6–8 weeks before the last frost. Or sow outside, after the last frost, directly where they are to grow. Seeds need light to germinate; gently press seeds onto surface and do not cover with medium.

Germination Temperature: 70°F

Days to Germination: 7–10

Growing On Temperature: 55–60°F

Garden Planting: After the last frost, plant in full sun in well-drained, sandy soil. Space 8–12 inches apart.

Description: Small white disk-shaped flowers with yellow centers bloom profusely in July and August. Can become invasive and weedy. This is an excellent flower for drying. Height: 2–2½ feet.

Chrysanthemum × morifolium 'Autumn Glory' hybrid mix

CHRYSANTHEMUM PTARMICIFLORUM (CINERARIA MARITIMA, SENECIO CINERARIA)

Annual (half-hardy)
Silver lace, dusty miller

Sowing Directions: Sow seed early since plants are slow growing. Sow indoors 9–12 weeks before the last frost. Can be sown outdoors, after the last frost, directly where they are to grow.

Germination Temperature: 70–75°F

Days to Germination: 10–15

Growing On Temperature: 65°F

Chrysanthemum ptarmiciflorum 'Silver Lace'

Garden Planting: Plant in full sun in any well-drained to dry soil; it likes drier conditions. Space 12 inches apart.

Description: Produces ornamental bluish green or silver-gray, finely dissected velvety foliage. The small yellow and white flowers are usually removed. Use for edging, bedding, containers, and borders. Height: 8–36 inches.

Recommended Cultivars:

Chrysanthemum ptarmiciflorum 'Silver Lace': Very finely cut foliage. Height: 7 inches.

Cineraria maritima 'Silverdust': Finely dissected, compact foliage. Height: 9 inches.

CHRYSANTHEMUM × SUPERBUM (C. MAXIMUM)

Perennial
Shasta daisy
Zones: 5–9

Sowing Directions: Sow seeds indoors 8 weeks before the last frost. Can be sown outdoors, after the first frost, where they are to grow. Seeds need light to germinate; gently press onto surface and do not cover with medium.

Germination Temperature: 65–70°F

Days to Germination: 9–12

Growing On Temperature: 55–60°F

Garden Planting: Plant in full sun in well-drained, moist, fertile soil. In southern climates, plant in part shade. Pinch plants in early summer for a bushier habit. Space 12–24 inches apart.

Description: This plant has a rounded habit and produces the classic daisy flowers with white rays and yellow centers that bloom from early summer to first frost. The flowers are very long lasting when cut. Height: 1–3 feet.

Recommended Cultivars:

'She Loves Me': Early-blooming (from late spring) variety with huge (6-inch) white flowers. Height: 3½ feet.

'White Knight' hybrid: Produces large (4-inch) white flowers on 20-inch stems. Height: 2 feet.

CIRSIUM JAPONICUM

Grow as an annual
Thistle, Japanese thistle, plume thistle

Sowing Directions: Sow seeds indoors 6–8 weeks before the last frost. Easy to sow outdoors, after the last frost, directly where they are to grow.

TIP:
In areas with mild winters, mulch silver lace well through the cold months. It may come up again the following year.

Germination Temperature: 70–75°F

Days to Germinate: 7–14

Growing On Temperature: 55–58°F

Garden Planting: Plant in full sun in well-drained soil. This plant will even tolerate poor, dry soil. Cut back after frost. Space 10–12 inches apart.

Description: Dark green spiny foliage with brilliant, richly colored red thistlelike flowers that appear in late summer and continue into fall. Used as a dried or fresh flower. Height: 2½ feet.

CLARKIA AMOENA

Annual
Godetia, Rocky Mountain garland

Sowing Directions: Because this plant does not transplant well, sow seeds outdoors, after the last frost, directly where they are to grow. In frost-free areas, sow seeds in the fall for early spring blooms. Seeds need light to germinate; cover with just a dusting of soil and keep well moistened.

Days to Germination: 5–10

Garden Planting: Plant in full sun to part shade in any well-drained soil. Thin seedlings to 10–14 inches apart.

Clarkia amoena

Description: Gorgeous native wildflower with large velvety red, pink, salmon, lavender, and bicolor single or double flowers that blooms in late spring and early summer. Thrives, with extended blooms, in cooler weather.

CLEMATIS

Perennial
C. × jackmanii: Jackman clematis (Zones 4–8); *C. montana*: anemone clematis (Zones 5–9); *C. Viticella*: Italian clematis (Zones 4–8)
Clematis, virgin's bower

Sowing Directions: For sowing indoors, about 12 weeks before the last frost place seeds in a moistened medium and put in a refrigerator for 3–4 weeks (40–42°F). After the cold treatment, place containers in a spot where the potting mix will warm to 80–85°F. Can be sown outdoors, in the late fall for spring germination, directly where they are to grow.

Germination Temperature: 80–85°F

Days to Germination 29–60 (sometimes longer)

Growing On Temperature: 70–75°F

Garden Planting: Plant in full sun to part shade in moist, rich, well-drained alkaline (limey) soil. Keep it well watered throughout the growing season. Space 5 feet apart.

Description: A vigorous vine that displays showy, ornamental flowers.

C. × jackmanii: Large (6-inch) dark purple, satiny flowers with white centers that bloom in early to late summer. Height: 10 feet.

TIP:

If growing clematis in full sun, make sure to add humus and mulch to the soil because it likes "cool feet" to grow well. Provide a trellis or other support, or grow clematis near shrubs and allow the vines to grow in and around them. Make sure the support structure is no more than 18 inches from the base of the plant.

C. montana: Vigorously growing vine that bears a profusion of 2-inch pink and lilac flowers that bloom in midspring to early summer. Height: 20–30 feet.

C. viticella: Longer-blooming species with 2–3-inch purple flowers that bloom from mid to late summer to late summer into early fall. Height: 12 feet.

CLEOME HASSLERANA (C. SPINOSA)

Annual
Cleome, spider flower

Sowing Directions: Sow indoors 6 weeks before the last frost. Seeds are most easily sown outdoors, after the last frost, directly where they are to grow. Seeds need light to germinate; cover very lightly with medium.

Cleome spinosa 'Violet Queen'

Germination Temperature: 70–75°F

Days to Germination: 10–14

Growing On Temperature: 55–60°F

Garden Planting: Plant in full sun in any well-drained soil. Space 3 feet apart.

Description: This unusual, wild, and rangy looking plant is drought- and heat-tolerant. It sports pink, laven-der, and white flowers that look like long whiskers; the stems are thorny. Height: 3–6 feet.

Recommended Cultivars: 'Queen' series—'Violet Queen', 'Rose Queen', 'Helen Campbell' (white), and 'Queen' mix.

COLEUS × HYBRIDUS

Annual
Coleus

Sowing Directions: For planting outdoors, sow seed indoors 8–10 weeks before the last frost. Can be sown outdoors directly where they are to grow. For a houseplant, sow anytime. Seeds need light to germinate; do not cover with medium.

Germination Temperature: 70–75°F

Days to Germination: 10–14

Growing On Temperature: 65–70°F

Garden Planting: Plant in shade to part shade in well-drained, rich, moist soil. Keep well watered during the growing season. Space plants 1–1½ feet apart.

Coleus 'Wizard' mix

Description: Grown for its richly colored ornamental foliage that is usually multicolored with shades of red,

TIP:
Mature plants will readily self-sow to the point of invasiveness, so allow enough garden space and weed out unwanted self-sown seedlings as they come up.

TIP:
Pinch off flowers to prolong the life of the plant. Pinching also encourages bushy growth.

pink, green, yellow, and creamy white. Height: 1–2 feet.

Recommended Cultivar: 'Wizard' mix is a large-leaved variety in a mixture of bright colors with a bushy habit that doesn't need pinching.

CONSOLIDA AMBIGUA

Annual
Larkspur, rocket larkspur, annual delphinium

Sowing Directions: Sow seeds indoors 6–8 weeks before the last frost. Or sow outdoors, after the last frost, directly where they are to grow. In frost-free areas, sow again in the fall for late winter and early spring blooms. Always use fresh seed, as it does not remain viable for long.

Germination Temperature: 55–65°F

Days to Germination: 10–20

Growing On Temperature: 50–55°F

Transplant after: 28–35 days

Garden Planting: Plant in full sun in rich, well-drained, alkaline soil. Space 12–18 inches apart.

Description: An easy plant to grow from seed, it displays showy spikes of red, purple, pink, and white flowers that resemble delphiniums. Height: 2–4 feet.

COREOPSIS GRANDI-FLORA, C. VERTICILLATA

Perennial
Coreopsis, tickseed
Zones: 4–9

Sowing Directions: Sow seeds indoors 8 weeks before the last frost.

Can be sown outdoors, in fall or early spring, directly where they are to grow. If sown early enough, plants grown from seed should produce flowers the first year. Seeds need light to germinate.

Germination Temperature: 65–70°F

Days to Germination: 20–25

Growing On Temperature: 55°F

Garden Planting: Plant in full sun in well-drained, sandy soil. Deadhead flowers to prolong bloom time. Space 12–18 inches apart.

Description: One of the easiest perennials to grow, it bears bright yellow daisylike flowers that bloom in early to mid and late summer (June and July). Light green foliage is finely cut. Height: 20–36 inches.

Recommended Cultivar:

'Early Sunrise' has radiant, semi-double yellow flowers that bloom just 11 weeks after indoor sowing and continue

Coreopsis grandifora 'Early Sunrise'

blooming all summer long. Height: 2 feet.

'Moonbeam' has light yellow flowers and reaches 18–24 inches in height.

'Zagreb' has bright yellow flowers and frows 12–18 inches in height.

CORTADERIA SELLOANA

Perennial
Pampas grass
Zones: 7–11

Sowing Directions: Sow seeds indoors 6–8 weeks before the last frost. Or sow outdoors, after the last frost, directly where they are to grow. Keep seed bed moist at all times. Seeds need light to germinate; do not cover with medium.

Germination Temperature: 65–75°F

Days to Germination: 20–27

Growing On Temperature: 60–65°F

Garden planting: Plant in full sun in well-drained, rich soil. North of Zone 7 cut plant back in the fall and dig up the roots. Store in a cool, dry location and replant in the following spring. In Zone 7 and south leave uncut during the winter for ornamental interest. Space 5–6 feet apart.

Description: A vigorously growing ornamental grass that produces striking 2–3-foot-high white feathery plumes. Height: 10–12 feet.

CORNFLOWER. SEE *CENTAUREA CYANUS*

COSMOS

Annual
C. bipinnatus, C. sulphureus
Cosmos

Sowing Directions: Sow seeds indoors 4–5 weeks before the last frost. Can be sown outdoors, after the last frost, directly where plants are to grow. Seeds are very fast growing.

Germination Temperature: 70°F

Days to Germination: 5–7

Growing On Temperature: 65°F

Garden Planting: Plant in full sun in well-drained soil. Will tolerate poor soils. Space 8–10 inches apart.

Description: These heat- and drought-resistant species have light, airy foliage and silky red, orange, pink, and white flowers that bloom from midsummer to frost. Height: 36–48 inches *(C. bipinnatus)*; 16–36 inches *(C. sulphureus)*.

Recommended Cultivars:

'Bright Lights' mix flowers come in yellow, lemon, gold, and orange. Height: 3 feet.

'Versailles' series plants have unusually beautiful flowers in pink, carmine, rose, and white. Height: 3–4 feet.

CUPHEA IGNEA

Annual
Cigar flower, cigar plant, firecracker plant

Sowing Directions: For best results, sow seeds indoors 6–8 weeks before

TIP:
Mature plants self-sow readily and make a great choice for adding bright color to any tough site.

TIP:

For a house-plant, grow cyclamen in a cool location (55–60°F). It thrives in high humidity, so mist the foliage several times a week. After the flowers die back, place the container in a shady site and cut back on water so the plant can rest. Begin watering again in late summer and fall for winter blooms.

the last frost. Can be sown outdoors directly where they are to grow. Seeds need light to germinate; do not cover with medium.

Germination Temperature: 70–75°F

Days to Germination: 7–14

Growing On Temperature: 65–70°F

Garden Planting: Plant in full sun to light shade in well-drained soil. Does not like dry conditions, so keep well watered. Space 12–14 inches apart.

Description: Has a round, compact habit. The bright red tubular flowers with their black-and-white tips resemble lit cigars. Height: 12 inches.

Cyclamen

CYCLAMEN PERSICUM

Perennial
Cyclamen
Zones: 9–11

Sowing Directions: Soak seeds overnight in warm (70–80°F) water before sowing. Seed needs total darkness to germinate; sow seed 1/2 inch below the surface. Spread a layer of sphagnum peat moss over the surface to hold in moisture; the seeds will be in the container for 6–18 months.

NOTE: *For Christmas and/or winter blooms, sow seed in early summer—it usually takes about 18 months for plants grown from seed to flower.*

Germination Temperature: 60°F

Days to Germination: 28–50

Growing On Temperature: 55°F

Garden Planting: In Zones 9–11, plant in the fall in part shade in rich, moist, well-drained soil for winter blooms. Space 8–12 inches apart.

Description: Very similar to *Cyclamen* species (see below), but the foliage has silver markings and the flowers can be single, double, or ruffled in red, pink, rose, light orange, white, purple, and lilac. Plants bloom during the winter when grown indoors.

Cyclamen 'Exotic' hybrid mix

CYCLAMEN SPECIES

Perennial
Hardy cyclamen
Zones: 5–7

Sowing Directions: Sow seeds indoors in late summer. Overwinter the containers in a cold frame, cool greenhouse, or other protected environment such as a porch. Seeds need darkness to germinate; cover completely with medium.

Germination Temperature: 55–60°F

Days to Germination: 21–28

Growing On Temperature: 55–60°F

Garden Planting: Plant in the spring, when the ground can be worked, in well-drained, rich soil. Responds well to heavy watering during dry periods. Space 6–8 inches apart.

Description: Charming low-growing plant whose flowers appear like tapestry on a green carpet of lawn and woodland grasses. Resplendent pink, light purple to red flowers bloom in late summer and fall and gently nod above dark green to bronzy purple heart-shaped foliage. *C. coum* is the exception, blooming in early spring as a cheerful sight after a long, colorless winter. Height: 3–4 inches.

CYNOGLOSSUM AMABILE

Annual
Chinese forget-me-not, hound's-tongue

Sowing Directions: Sow seeds indoors 6–8 weeks before the last frost. Or sow seeds directly outdoors as soon as the soil can be worked—even before the last frost date. Seeds need darkness to germinate; cover completely.

Germination Temperature: 65–70°F

Days to Germination: 5–10

Growing On Temperature: 60°F

Garden Planting: Plant in full sun in any well-drained soil. Space 12 inches apart.

Description: An old-fashioned-looking plant with arching branches that bear delicate, dainty, single light blue, pink, or white flowers that bloom from early to late summer, sometimes into early fall. Flowers are lightly fragrant. Height: 18–24 inches.

DAHLIA

Perennial (tender)
Dahlia
Zones: 9–11

Sowing Directions: Sow indoors 8 weeks before the last frost. It is important to keep soil temperature constant during germination.

Germination Temperature: 60–65°F

Days to Germination: 5–10

Growing On Temperature: 55–60°F

Garden Planting: Plant in full sun in fertile, well-drained soil. Keep plants well-watered and fertilize regularly or apply a slow-release fertilizer at the beginning of the growing season. Taller varieties (those over 3 feet) should be staked for extra support. Space 2 feet apart, depending on variety.

Description: Hybrids range from statuesque, gorgeous garden queens to marigoldlike lollipop flowers. Flowers can be single, double, spidery, pompom, or quilled and in almost any color, including red, pink, white, orange, yellow, and

TIP:
After the first fall frost, dahlia tubers can be dug up, cleaned, and stored during the winter in a cool, dark, dry place. Make sure the tubers don't dry out completely; check periodically. Replant in the spring after the last frost.

maroon; they bloom from midsummer through frost. From seed, some varieties can reach up to 4 feet in the first season. Height: 1–5 feet.

Recommended Cultivar: 'Rigoletto' mix has large double and semidouble flowers in white, yellow, pink, orange, red, and rosy purple. Early-blooming variety. Height: 13 inches.

TIP:
Dahlias grown from seed cost a fraction of what the tubers cost. Since they are so easy to grow, save some money and start plants from seed the next time you want some of these flowers in your garden.

Dahlia 'Cactus Rose'

DELPHINIUM ELATUM

Perennial
Delphinium
Zones: 3–7

Sowing Directions: Sow indoors 2 months before the last frost. Plants grown from seed sown this early will bloom the first year. Can also be sown outdoors, in late spring and early summer, directly where they are to grow. Plants grown from seed sown directly will not bloom until the following year. Always use fresh seed, as it will not remain viable for long. Seeds need darkness to germinate; cover completely with medium.

Germination Temperature: 65–75°F

Days to Germination: 12–21

Growing On Temperature: 50–55°F

Garden Planting: Plant in full sun in rich, well-drained, alkaline, deeply worked soil. Keep well-watered and fertilized. Prefers cool climates, so extra watering is particularly important during hot, dry spells. Cut back after bloom; stake taller varieties. Space 2 feet apart.

Description: Striking spikes of soft blue, white, pink, lilac, and yellow flowers bloom in June and July. A must for any Anglophile's timeless cottage garden. Height: 2–6 feet (in bloom).

Delphinium elatum 'Fantasia', a semidwarf

Recommended Cultivar: 'Fantasia' mix has strong plants with heavenly light blue, dark blue, and white flowers. Height: 27 inches.

Dianthus

DIANTHUS CARYO-PHYLLUS, D. CHINENSIS

Annual
Dianthus, china pinks, annual pinks, carnation

Sowing Directions: Sow indoors 6–8 weeks before the last frost. Or sow directly outdoors after the last frost. Plants grown from seed take about 3 months to bloom.

Germination Temperature: 60–70°F

Days to Germination: 7–14 days

Growing On Temperature: 55–60°F

Garden Planting: Plant in full sun in fertile, well-drained, slightly alkaline soil. Deadhead spent flowers to prolong blooming. Space 10–12 inches apart.

Description: Small, compact plants with fringed fragrant pink, red, white, or lavender single or double flowers. In areas with mild winters, may overwinter or self-sow to emerge and grow the following year.

D. Caryophyllus: A tender perennial grown as an annual that blooms in midsummer. Height: 1–2 feet.

D. chinensis: A low-growing plant that blooms in early summer through to fall. Height: 6–15 inches.

Recommended Cultivar: 'Floral Lace' series has exceptionally heat-tolerant hybrids.

DIANTHUS DELTOIDES, D. PLUMARIOUS, D. BARBATUS D. GRATIA-NOPOLITANUS, D. × ALLWOODII

Biennial; perennial
Zones: 4–8

Sowing Directions: For best results, sow perennials indoors 8–10 weeks before the last frost. Or sow outdoors, after the last frost, directly where they are to grow. The biennial *D. barbatus* should be sown indoors 6–8 weeks before the last frost.

Germination Temperature: 60–70°F

Days to Germination: 14–21

Growing On Temperature: 55–60°F

Garden Planting: Plant in well-drained, loose, slightly alkaline soil. Does not thrive in hot, dry conditions. Space 12 inches apart.

Description: These species have fringed single or double flowers with blue-green to grayish flowers that begin blooming in late spring through early summer with white, pink, red, and bicolored flowers. Height: 6–24 inches.

D. × Allwoodii is a vigorous growing and compact hybrid cross. Height: 8–18 inches tall.

D. barbatus (sweet William) has fragrant flowers and self-sows. Height: 24 inches.

D. deltoides (maiden pink) is evergreen tufted-shaped with small white, pink and red flowers. Height: 6–12 inches.

D. gratianopolitanus (cheddar pinks) has a low-growing, spreading habit. Height: 6 inches.

D. plumarius (cottage pink, garden pink, grass pink) has red, pink, white, or bicolored flowers with dark centers. Makes a good rock garden or edging plant. Height: 12 inches.

Recommended Cultivars:

'Spring Beauty' has fragrant semi-double flowers with foliage

TIP:
It is best to buy dianthus hybrids that are heat tolerant—a must for gardeners living in warmer climates.

that forms a dense mat. Height: 12–18 inches.

'Zing Rose' has vivid red rose flowers that bloom all summer with grassy foliage. Height: 6 inches.

Dianthus 'Spring Beauty'

DICENTRA SPECTABILIS, D. EXIMIA

Perennial
Bleeding heart
Zones: 3–8

Sowing Directions: For indoor sowing, 14–16 weeks before the last frost place seeds in a moistened medium and refrigerate (below 40°F) for 6–8 weeks. Begin to germinate about 8 weeks before the last frost. For outdoor sowing (which is easier), sow seeds outdoors, in late summer to late fall, directly where they are to grow; they will germinate the following spring.

Germination Temperature: 55–60°F

Days to Germination: 30–45 or more

Growing On Temperature: 55–60°F

Garden Planting: Plant in part shade in rich, moist, well-drained soil. Will tolerate full sun if kept very well watered. Space 18–24 inches apart.

Description: Sometimes called the "living valentine plant," it has an old-fashioned charm with its dangling pink and/or white heart-shaped flowers that bloom on arching branches of graceful dissected foliage.

D. eximia has a clumped habit; blooms begin in early summer and can continue until fall. Height: 18 inches.

D. spectabilis has showy, large, bright flowers that bloom from late spring to early summer. Height: 2½ feet.

Dicentra spectabilis

DIDISCUS. SEE TRACHYMENE

DIGITALIS

Perennial; biennial
D. grandiflora, D. × mertonensis
(Zones 4–8), D. purpurea
Foxglove, finger flower, fairy
glove

Sowing Directions: Sow indoors 6–8 weeks before the last frost. It is easier to sow outdoors, in summer or early fall, directly where they are to grow for bloom the following year. Seeds need light to germinate; gently press seeds onto surface and keep seed beds moist.

Germination Temperature: 65–70°F

Days to Germination: 10–15

Growing On Temperature: 65°F

Garden Planting: Plant in full sun to part shade in fertile, well-drained slightly acidic soil. Space 18–24 inches apart.

Digitalis species

Description: A tall-growing, dramatic yet old-fashioned, cottage garden plant. Its erect spikes of long bell-shaped pink, white, buttery yellow, or purple flowers with spotted throats bloom in early to midsummer (June–July). Height: 2–5 feet.

D. *grandiflora* has yellow flowers that bloom from late spring to early summer. Prefers part shade. Height: 2 feet.

D. × *mertonensis* has red to mauve flowers. Height: 2½ feet.

D. *purpurea* is the classic biennial and readily self-sows. Height: 2–5 feet.

DIMORPHOTHECA SINUATA

Annual
Cape marigold, African daisy, star of the veldt

Sowing Directions: Best to sow indoors 5–6 weeks before the last frost. Can be sown outdoors, after the last frost, directly where they are to grow. In Zones 9–11, they can be sown directly outdoors either in late fall for winter blooms or in winter for early spring blooms. Always use fresh seed as it is not long-lived.

Germination Temperature: 60–70°F

Days to Germination: 10–15

Growing On Temperature: 55–60°F

Garden Planting: Plant in full sun in rich, well-drained soil. Prefers a rich soil, but will grow in poorer sandy or dry soils. Keep well-watered in dry conditions. Prefers cool condi-

Dimorphotheca sinuata

TIP:
Established plants will self-sow. Either thin out the seedlings or, in the fall, transplant them to where you would like them to grow. If you cut foxglove back after it blooms, you will prevent it from self-sowing and, as a bonus, you may get a second bloom.

tions (especially cool nights) and doesn't thrive in hot, humid conditions. Space 10–12 inches apart.

Description: The bright clear yellow, white, pink, and yellow daisylike flowers have dark centers and bloom in late spring to fall. Use for borders, bedding, and cut flowers. Height: 12–15 inches.

DODECATHEON MEADIA

Perennial
Shooting star, American cowslip, prairie pointer
Zones: 4–8

Sowing Directions: For indoor sowing, 10–12 weeks before the last frost, place seeds in a plastic bag with a moistened medium and refrigerate for 3–4 weeks. It is easier to sow directly outdoors in late fall or early winter while the soil can still be worked lightly. Seeds need light to germinate; press gently onto the surface.

Indoor Germination Temperature: 60–70°F

Days to Germination: 25–30 or more

Growing On Temperature: 55–60°F

Garden Planting: Plant in part to full shade in rich, moist, well-drained, acidic soil. Keep well watered during the growing season. It is excellent for naturalizing shady areas. Foliage will die back in the full heat of summer. Space 12–18 inches apart.

Description: The distinctive flowers resemble flaming darts that look as if they had been caught in midair while streaming across a night sky. The

deep rosy flowers with pointed dark centers appear from spring to early summer. Height: 6–20 inches.

DOLICHOS LABLAB

Annual
Hyacinth bean, lablab

Sowing Directions: Soak seeds overnight in warm water before sowing. To get a jump on growing season, sow seeds indoors 6–8 weeks before the last frost into individual pots. Can be sown outdoors, just after the last frost, directly where they are to grow. Cover seeds lightly.

Indoor Germination Temperature: 70–75°F

Days to Germination: 15–30

Growing On Temperature: 65°F

Garden Planting: Plant in full sun in well-drained soil. Transplant seedlings after the soil has warmed

Dolichos lablab

up in late spring or early summer; they will not be active in cool soils but will grow quickly when the soil temperature climbs. Support plants with a fence, trellis, or other structure. Space 24 inches apart.

Description: Easy to grow and maintain, this vine displays showy light purple, white, or yellow pealike flowers that bloom all summer. After flowering, long (2–8-inch), leathery purple seedpods develop. Height: 6 feet.

DORONICUM CORDATUM

Perennial
Leopard's-bane
Zones: 4–8

Sowing Directions: Sow seeds indoors 8–10 weeks before the last frost. Can be sown directly outdoors in early fall or early spring (up to 2 months before first frost). Seed needs light to germinate; press onto surface.

Indoor Germination Temperature: 70°F

Days to Germination: 14–21

Growing On Temperature: 55–60°F

Garden Planting: Plant in part shade in moist, well-drained, rich soil. Prefers cool growing conditions; keep well watered during hot spells. Space 12–14 inches apart.

Description: Bright yellow daisylike flowers with tufted centers bloom in late spring to early summer. Use for bedding, borders, cut flower, or rock gardens. Height: 18–24 inches.

DUSTY MILLER. SEE *CHRYSANTHEMUM PTARMICIFLORUM*; *CINERARIA MARITIMA*

E

ECHINACEA PURPUREA

Perennial
Purple coneflower, hedgehog coneflower, purple rudbeckia
Zones: 3–9

Sowing Directions: Sow seeds indoors 8–10 weeks before the last frost. Can be sown directly outdoors as soon as soil can be worked in early spring and anytime up to 2 months before the first frost. If sown early enough indoors, plants will bloom the first year, otherwise expect bloom in the second year from seed. Seeds need darkness to germinate.

Germination Temperature: 70–75°F

Days to Germination: 8–21

Growing On Temperature: 50–55°F

Garden Planting: Plant in full sun in well-drained soil. While not too fussy about soil, it prefers acidic soil. Space 18–24 inches apart.

Description: This heat- and drought-tolerate native perennial provides beauty and attracts butterflies to any garden. The incomparable rosy purple daisylike flowers, which bloom

from midsummer to fall, have petals that droop downward from brownish orange, thistlelike centers that have a mesmerizing iridescent quality to them. A white variety is also available. Height: 2–3½ feet.

ECHINOPS RITRO

Perennial
Globe thistle
Zones: 3–9

Sowing Directions: Sow seed indoors 2–4 weeks before the last frost. Can be sown directly outdoors anytime after the last frost up to 2 months before the first fall frost. Seeds need light to germinate; cover sparsely with medium.

Indoor Germination Temperature: 65–75°F

Days to Germination: 15–21

Growing On Temperature: 55–60°F

Garden Planting: Plant in full sun to light shade in well-drained soil. Will grow in dry soils. Tall specimens may need staking. Space 15–20 inches apart.

Description: The unusual iridescent, prickly bluish globe flowers make an

Echinops ritro

excellent contrast in a border planting, especially next to bright red- and yellow-flowering plants. Use in the middle to the back of borders. Height: 2–4 feet.

ERIGERON

Perennial
Fleabane, daisy fleabane
Zones: 4–10

Sowing Directions: Sow seeds indoors 8–10 weeks before the last frost. Can be sown directly outdoors in early spring or early fall. Lightly cover seeds with medium.

Indoor Germination Temperature: 50–55°F

Days to Germination: 15–25

Growing On Temperature: 50–55°F

Garden Planting: Plant in full sun in well-drained soil. Space 12 inches apart.

Description: Abundant, showy, richly lavender, white, pink, and orange asterlike flowers with bright yellow centers bloom profusely from early summer to early fall. Use in borders, in rock gardens, or as a cut flower. Height: 1½–2 feet.

Recommended Cultivar: 'Azure Fairy' is very easy to grow and has 1½-inch lavender flowers. Height: 1½ feet.

ESCHSCHOLZIA CALIFORNICA

Annual
California poppy

Sowing Directions: For indoor sowing, start seeds in peat pots that can be directly transplanted into the garden 2–3 weeks before the last frost. Easier to sow outdoors, after the last frost, directly where they are to grow. Lightly cover seeds with medium.

Indoor Germination Temperature: 60–65°F

Days to Germination: 14–21

Growing On Temperature: 55–60°F

Garden Planting: Plant in full sun in any well-drained soil. It will tolerate dry soils, but extra watering promotes increased flowering. Space 8–10 inches apart.

Description: Lively bright red, orange, white, and yellow poppylike flowers cheer up the garden from midspring through fall. Foliage is light and feathery. Mature plants will usually self-sow. Excellent for

Eschscholzia californica 'Ballerina' mix

borders, rock gardens, and container plantings. Height: 1–2 feet.

Recommended Cultivar: 'Ballerina' mix has luminous (3½-inch) yellow, rose, orange, scarlet, and carmine flowers in solids and bicolors. Very easy to grow. Height: 10 inches.

EUPATORIUM

Perennial
E. maculatum, E. purpureum
Joe-pye weed, boneset, mist flower
Zones: 3–8

Sowing Directions: For indoor sowing, start in late winter. Place seeds in individual peat pots in a moistened medium and refrigerate(40–45°F) for 2–3 months; then germinate at 55–60°F. It is easier to sow outdoors, in the fall or early spring, directly where they are to grow. Plants started from seed take 2 years to bloom; thus they are usually propagated through division. Use fresh seed as it is not long lived. Lightly cover seeds with medium.

Indoor Germination Temperature: 55–60°F

Days to Germination: 30–90

Growing On Temperature: 55°F

Garden Planting: Plant in full sun to very light shade in rich, well-drained, acidic soil. Taller specimens may need staking. Space 2–3 feet apart.

Description: A tall, bright, and ruggedly handsome native plant with clusters of rounded deep red to dark lavender-purple and maroon flowers that bloom in late summer through midfall. It is very tough and readily withstands heat but should

be kept moist in drought; a must-have for naturalized meadow plantings. Use for the back of the border. Height: 4–6 feet.

Euphorbia

EUPHORBIA CYATHOPHORA, E. HETEROPHYLLA, E. MARGINATA, E. VARIEGATA

Annual
Annual poinsettia, crown of thorns, fire-on-the-mountain, Mexican fire plant, snow-on-the-mountain, spurge

Sowing Directions: Sow indoors 6–8 weeks before the last frost. In southern zones only, can be sown outdoors after the last frost. Lightly cover seeds.

Indoor Germination Temperature: 70–75°F

Days to Germination: 10–25

Growing On Temperature: 60°F

Garden Planting: Plant in full sun or partial shade in any well-drained soil. Although it is heat- and drought-resistant, keep it watered during hot, dry spells for best performance. Space 10–12 inches apart.

NOTE: *Some people develop a rash from the milky substance in the stems of spurges.*

Description: Colorful bright red or white upper foliage develops above lower bright green foliage. Small flowers bloom during midsummer. Height: 18–24 inches.

Euphorbia species, snow-on-the-mountain

EUPHORBIA MYRSINITES, E. POLYCHROMA (E. EPITHYMOIDES)

Perennial
Spurge, cushion spurge, myrtle euphorbia
Zones: 4–9

Sowing Directions: For indoor sowing, place seeds in a plastic bag with moistened medium and refrigerate for 4–6 weeks before bringing out to germinate at warmer temperatures. If there isn't enough time to subject the seeds to the cold treatment, sow untreated seeds 6–8 weeks before the last frost. For outdoor sowing, sow after the last frost directly where they are to grow. Cover seeds with a soil layer two to three times their width.

Indoor Germination Temperature: 65–70°F

Days to Germination: 8–21

Growing On Temperature: 50–55°F

Garden Planting: Plant in full sun in any well-drained, dry soil. Water well, but allow soil to dry out between waterings. Space 12 inches apart.

Description: Succulent-type plant with distinctive grayish blue foliage. Use as ground cover or edging and in borders or rock gardens.

E. myrsinites produces whorls of foliage and yellow bract flowers in the spring. Height: 8–10 inches.

E. polychroma is grown mainly for its spicy crimson-colored fall foliage. Height: 12–16 inches.

EUSTOMA GRANDIFLORUM

Annual
Lisianthus, prairie gentian

Sowing Directions: Sow seeds indoors 10–12 weeks before the last frost into individual peat pots that can be transplanted directly to the garden; the plants don't respond well to transplanting. In southern zones where the growing season is very long, it can be sown directly outdoors. Pinch seedlings when they are about 6 inches tall to encourage branching and increase flowering.

Seeds need light to germinate; gently press onto surface and do not cover with medium.

Indoor Germination Temperature: 65–75°F

Days to Germination: 10–20

Growing On Temperature: 55–60°F

Garden Planting: Plant in full sun in moist, well-drained, dry soil. Space 12 inches apart.

Description: Produces a swirling delight of upright stems with lovely single and double blue, purple, pink, and creamy white silky flowers that are long lasting when cut. Makes a gorgeous mass planting in a border. Height: 18–36 inches.

Recommended Cultivar: 'The Blue Rose' has romantic roselike blue flowers that last a full 2 weeks when cut. Height: 18–22 inches.

EVENING PRIMROSE. SEE *OENOTHERA MISSOURENSIS*

FELICIA

Annual
F. amelloides, F. Bergerana
Blue daisy, blue marguerite, kingfisher daisy

Sowing Directions: For best results for indoor sowing, place seeds in a

plastic bag with moistened medium and place in a refrigerator for 2–3 weeks. Sow cold-treated seeds 6–8 weeks before the last frost. Can sow outdoors, after the last frost, directly where they are to grow. Lightly cover seeds with medium.

Indoor Germination Temperature: 60–65°F

Days to Germination: 30–35

Growing On Temperature: 55–60°F

Garden Planting: Plant in full sun in dryish, well-drained, rich soil. Space 12–16 inches apart.

Description: A cheerfully blue daisy-like flower with prominent yellow centers that bloom from midsummer to frost. Resembles the Swan River daisy (*Brachycome iberidifolia*). Use for edgings, borders, or container plantings.

F. amelloides flowers from midsummer to frost. Height: 2–3 feet.

F. Bergerana has a low-growing, mat-forming, spreading habit. Height: 4–6 inches.

FESTUCA OVINA

Perennial
Blue fescue
Zones: 4–9

Sowing Directions: For best results, sow seeds outdoors, in early spring to early summer, directly where they are to grow. Press seeds onto surface; they do not need to be covered with medium.

Germination Temperature: 70–75°F

Days to Germination: 21–25

Growing On Temperature: 55–60°F

Garden Planting: Plant in full sun to part shade. Blue color will be best if grown in dryish soil and full sun. Space 14–18 inches apart.

Festuca ovina

Description: A tufted ornamental grass with silvery blue to bluish gray foliage. Makes an excellent seashore planting as it is quite salt-tolerant. Use in masses for edgings or contrast with flowering plants. Height: 8 –18 inches.

Recommended Cultivar: 'Elijah Blue' foliage has a pronounced steely blue effect. Height: 10 inches.

FILIPENDULA

Perennial
F. purpurea: Japanese meadowsweet, *F. rubra*: queen of the prairie, *F. ulmaria*: queen of the meadow
Zones: 3–8

Sowing Directions: For indoor sowing, sow seeds 8–10 weeks before the last frost. For outdoor sowing, sow into containers in fall and sink into the ground. Cover with a sheet of heavy plastic or glass. Water as needed throughout the winter and early spring. Transplant seedlings into larger containers and grow on until late summer or fall and transplant to their permanent spot. Plants grown from seed may not bloom until the second year. Plants are often propagated by division.

Indoor Germination Temperature: 55–60°F

Days to Germination: 25–90

Growing On Temperature: 55°F

Garden Planting: Plant in full sun. after the last frost, in moist, fertile, well-drained, slightly acidic soil. Taller-growing varieties should be staked for best effect. To increase winter interest, leave the dried flower head on the plants. Space 15–24 inches apart.

Description: A tall, graceful plant with light, frothy, feathery puffs of white and pink flowers and fernlike foliage. The flowers, which generally bloom in early to midsummer, are excellent for fresh or dried arrangements. Terrific for the back of the border and naturalized meadow plantings. Height: 3–8 feet.

F. purpurea produces pink flowers on sturdy stems that do not need staking. Height: 4 feet.

F. rubra has light pink flowers. Height: 6–8 feet.

F. Ulmaria has white flowers. Height: 5–6 feet under good growing conditions; otherwise 3–4 feet.

Recommended Cultivars (*F. Ulmaria*):

'Aurea' has notable golden foliage.

'Flore Pleno' has double white flowers.

'Variegata' has variegated foliage.

FORGET-ME-NOT. SEE *MYOSTOTIS SYLVATICA*

FOUR-O'CLOCK. SEE *MIRABILIS JALAPA*

FUCHSIA

Perennial; grow as an annual
Fuchsia, lady's eardrops
Zones: 10–11

Sowing Directions: For outdoor planting, sow seeds indoors 8–10 weeks before the last frost. For a houseplant, sow seeds anytime of the year. Seeds need light to germinate; do not cover with medium.

Fuchsia 'Evening Belle'

TIP:
Mature filipendula plants will self-sow. Gently dig up the young seedlings and transplant them to where you would like more plants.

TIP:
Dig up fuchsia before the first frost and grow as a houseplant for blooms throughout the winter. Or keep plants at about 40°F in low light and cut back on fertilizing and watering until the following spring.

TIP:
Fuchsias are most often used for container plantings. Make sure the containers and potting mix allow for good drainage. In addition to watering, mist the foliage several times a week, since fuchsias like humid growing conditions.

Indoor Germination Temperature: 70–75°F

Days to Germination: 21–30

Growing On Temperature: 65–70°F

Garden Planting: Plant in part to full shade in moist, light, well-drained soil. Keep fertilized and well watered throughout the growing season. For bushier plants and more blooms, pinch back in early summer. Space 2–2½ feet apart.

Description: A gorgeous flower that is hard to equal in garden-variety glamour. The flowers, which look like dangling earrings in silky jewel-like pink, red, purple, blue, orange, and white, will bloom throughout the summer. Height: 1–4 feet (higher in frost-free zones).

G

Gaillardia

GAILLARDIA PULCHELLA

Annual
Annual blanketflower, blanketflower, Indian blanketflower

Sowing Directions: For best results, sow seeds indoors 5–6 weeks before the last frost. Can be sown outdoors, after the last frost, directly where they are to grow. Seeds need light to germinate; press seeds lightly onto surface and do not cover with medium.

Indoor Germination Temperature: 70–75°F

Days to Germination: 5–15

Growing On Temperature: 55–60°F

Garden Planting: After the last frost, plant in full sun in any well-drained, sandy soil. Will tolerate dry or poor soils and benefits from an early summer application of fertilizer. Deadhead to extend the blooming period. Space 12–14 inches apart.

Description: The bright red, yellow, orange, and bronze single or double flowers bloom all summer through to frost and are long lasting when cut. A terrific, colorful, low-maintenance plant for edgings, beddings, and borders. Height: 6–30 inches.

Gaillardia pulchella 'Gaiety' mix

Recommended Cultivar: 'Gaiety' mix produces a multitude of claret, rose, orange, yellow, and maroon flowers; bicolors add to the howling mob of color and delight. Great for hot, dry growing conditions and/or poor soils. Height: 2 feet.

GAILLARDIA ×GRANDIFLORA

Perennial
Perennial blanketflower, blanketflower
Zones: 3–9

TIP:

If you live in a region with summer temperatures that climb above 90°F, this heat- and drought-tolerant plant is for you.

Sowing Directions: Sow seeds indoors 6–8 weeks before the last frost. Seeds need light to germinate; slightly press onto surface; do not cover with medium.

Indoor Germination Temperature: 70–75°F

Days to Germination: 5–15

Growing On Temperature: 55–60°F

Garden Planting: In late spring, after soil has warmed up a bit, plant in full sun in moist, slightly acidic, well-drained soil. It will tolerate almost any soil, including poor, dry conditions. Do not fertilize heavily; a little compost or manure in the late spring will do just fine. Space 12–16 inches apart.

Description: If you see this plant in bloom, it'll be easy to see the inspiration for some Native Americans' gorgeous blanket designs and colors. The bright red, yellow, bronzy brown, and orange daisylike flowers bloom all summer and fall, if cut back at the end of its initial flowering. Height: 14–36 inches.

Recommended Cultivar: 'Portola Giants' has scarlet, yellow, and gold-touched flowers whose colors glow like the colors of a fine sunset. Height: 2 1/2 feet.

GAURA LINDHEIMERI

Perennial; often grown as an annual
Gaura
Zones: 6–9

Sowing Directions: Sow seeds indoors 6–8 weeks before the last frost. Can be sown outdoors, after the last frost, directly where they are to grow. If grown as a perennial, sow outdoors in early spring or autumn. Seeds need light to germinate; press onto surface lightly and do not cover with medium.

Indoor Germination Temperature: 65–75°F

Days to Germination: 7–21

Growing On Temperature: 50–55°F

Garden Planting: Plant in midspring in full sun in any well-drained soil. Because it is very heat- and drought-tolerant, it will not need extra watering unless drought conditions prevail. Deadhead flowers to extend the blooming period, and cut back in late fall. Space 24–30 inches apart.

Description: This shrublike perennial has long flower stalks that bear white to pink flowers from early summer into the fall. Used in the back of borders and in wildflower plantings; makes a nice fresh-cut flower. Height: 5 feet.

Gaura Lindheimeri

GAZANIA RIGENS

Annual
Gazania, treasure flower

Sowing Directions: Sow indoors 6–8 weeks before the last frost. Can also be sown directly outdoors after the last frost. Seeds need darkness to germinate; lightly cover.

Indoor Germination Temperature: 70°F

Days to Germination: 10–14

Growing On Temperature: 55–60°F

Garden Planting: Plant in full sun in any well-drained soil. It prefers sandy soils and other dryish acidic soils. Be careful not to overwater, as it is susceptible to crown rot. Space 10–14 inches apart.

Description: The long-blooming, colorful, daisylike flowers come in sizzling hot shades of yellow, orange, red, pinks and striped bicolors. A terrific heat- and drought-tolerant plant. Use massed in beds or borders. Height: 8–12 inches.

Gazania rigens 'Pinata' hybrid mix

Recommended Cultivar: 'Pinata' hybrid mix produces large (4-inch) flowers in all of the above-listed colors and stripes. Height: 12 inches.

GERANIUM, ANNUAL. SEE *PELARGONIUM × HORTORUM*

GERANIUM

Perennial
Cranesbill, hardy geranium
Zones: 4–11

Sowing Directions: For indoor sowing, 10–11 weeks before the last frost, place seeds in a moistened medium and store in the refrigerator for 2–3 weeks. Sow the cold-treated seeds 8 weeks before the last frost. For outdoor sowing, in the fall or early spring, sow seeds into a container and sink it into the ground in a sheltered spot. Cover with thick, clear plastic or a pane of glass and water seedlings as necessary if it the soil looks too dry. Thin and transplant seedlings as they come up. Always use fresh seed, as it doesn't remain viable for long.

Indoor Germination Temperature: 65–75°F

Days to Germination: 10–90, depending on species

Growing On Temperature: 60–65°F

Transplant After: Seedlings are 3 inches tall

Garden Planting: After the last frost, plant in full sun to part shade in any well-drained acidic to slightly acidic soil. Make sure it receives extra water during hot, dry spells. Space 10–12 inches apart.

Description: Most species have a mounding to spreading habit with flowers in lovely soft shades of pink, purple, and white. Use in the border, as a ground cover, for container plantings, and as an informal edging. Height: 5 inches–2 feet, depending on the species.

Recommended Cultivars:

'Johnson's Blue' has lovely 2-inch violet-blue flowers borne on 15–18-inch stems.

G. *Endressii* 'Wargrave Pink' is low growing with abundant clear, bright pink flowers. Height: 12 inches.

GERBERA JAMESONII

Annual
Gerbera, Transvaal daisy, African daisy, veldt daisy, Barberton daisy

Sowing Directions: For best result, sow seeds indoors 10–12 weeks before the last frost. It is important to use fresh seed from a reputable company. Seeds need light to germinate; lightly press onto surface (placing the sharp end of the seed down) and do not cover with medium.

Indoor Germination Temperature: 70–75°F

Days to Germination: 15–25

Growing On Temperature: 65°F

Garden Planting: After the last frost, plant in full sun to part shade in moist, rich, well-drained, slightly acidic soil. Space 10–12 inches apart.

Description: These delightful daisy-like flowers come in a range of yellow, red, and pink colors and bloom all season. Most often used massed in borders, in container plantings, or for cut flowers. Height: 12–18 inches.

Gerbera Jamesonii 'California' mix

Recommended Cultivar:
'California' mix has large (3-inch) flowers borne on long, graceful stems. Flowers come in a range of yellow, orange, crimson, white, and pink. Height: 12 inches.

GEUM

Perennial
G. *Quellyon*: Zones 5–9, G. *reptans*, Zones 4–8
Geum, avens

Sowing Directions: Sow seeds indoors 8 weeks before the last frost. Or sow outdoors, up to 2 months before the first fall frost, directly where they are to grow. Can be sown into covered containers in the fall or early spring. Lightly cover seeds with medium.

Indoor Germination Temperature: 65–70°F

Days to Germination: 21–30

Growing On Temperature: 55–60°F

Garden Planting: Plant outdoors

anytime after midspring in full sun (to light shade in warmer zones) in rich, well-drained, slightly acidic soil. Water well during hot, dry periods if rainfall is not adequate. Deadhead spent flowers to prolong blooming time. Space 12–15 inches apart.

Description: The orange, red, or yellow 1–2-inch flowers bloom from late spring through early summer. The finely cut foliage gives it a light, airy appearance. For best effect, plant in masses in the border. Height: 6–36 inches.

GLADIOLUS × HORTULANUS

Perennial (tender bulb); grow as an annual
Gladiolus, glads
Zones: 9–11

> **NOTE:** *It is recommended that beginning gardeners buy gladiolus corms from a catalog or nursery center.*

Sowing Directions: For best results, sow seeds indoors 10–12 weeks before the last frost. Transplant seedlings several weeks after the last frost. For outdoor sowing, sow seeds directly where they are to grow. Thin seedlings to about 6 inches apart. Seeds will germinate and grow to form corms the first year; plants grown from seed take 2–3 years to bloom. Lightly cover seeds.

Indoor Germination Temperature: 70°F

Days to Germination: 20–40

Growing On Temperature: 60°F

Garden Planting: If grown as a perennial, plant seedlings in full sun in deeply worked, rich, well-drained, slightly acidic soil. When selecting a location, make sure there is good air circulation all around the plants. Must be well watered and fertilized throughout the growing season. Taller specimens should be staked. Space 6 inches apart.

Description: Like foxgloves and hollyhocks, these tall, erect, classic cottage garden flowers are showstoppers. The flowers, which come in almost all colors, bloom in the summer and, if planted in succession, into fall. Use in the back of the border. Height: 3–5 feet.

GLORIOSA SUPERBA

Perennial
Gloriosa lily
Zones: 9–11

Sowing Directions: Seeds can be sown indoors in late winter to early spring into individual peat pots to minimize transplant shock. In Zones 8–11, seed can be sown outdoors, in the spring, directly where they are to grow. Many gardeners recommend soaking the seeds in warm water overnight before sowing. Lightly cover seeds.

Indoor Germination Temperature: 70–75°F

Days to Germination: 21–30

Growing On Temperature: 70–75°F

Garden Planting: Plant outside after temperatures remain above 50°F. Plant in full sun in fertile, well-drained soil near some kind of

TIP:
If grown as an annual, plant corms in midspring and every 2 weeks throughout the spring for continuous summer blooms. Dig up gladiolus corms before the first frost and store in a cool, dry, frost-free place. Replant the following spring.

TIP:
It's easiest for beginning gardeners to buy tubers to plant.

support: a trellis, fence, or wall. Keep it well watered and fertilized during growing season. Space 12–18 inches apart.

Gloriosa superba

Description: This distinctive vining plant has exotic, spiderlike, recurved, red flowers with yellow edges and centers that bloom during the summer. Height: 3–6 feet.

GOMPHRENA GLOBOSA

Annual
Globe amaranth

Sowing Directions: Sow seeds indoors 6–8 weeks before the last frost. Can be sown outdoors, after the last frost, directly where they are to grow. Seeds need darkness to germinate; lightly cover with ¼ inch of medium.

Indoor Germination Temperature: 70–75°F

Days to Germination: 10–14

Growing On Temperature: 68–70°F

Garden Planting: After the last frost, plant in full sun in any well-drained soil. It prefers a slightly acidic soil. For bushier plants and increased flowering, pinch back young plants before they have begun to flower. Space 10–12 inches apart.

Description: This easy-to-grow plant produces spectacular bright balls of small cloverlike flowers in white, scarlet, violet, pink, and rose. The everlasting flowers are used in dried arrangements and for crafts. This very heat- and drought-tolerant plant can be massed in borders or used in beds, edgings, and containers. Height: 12–24 inches.

Gomphrena globosa
'Strawberry Fields'

Recommended Cultivar: 'Strawberry Fields' adds a unique strawberry red color from July to October and much charm to the garden. The name alone should win you over! Height: 24 inches.

TIP:
If growing gloriosa lily in cold climates, dig up the tubers before the first fall frost and store in a cool (50–55°F), dry place over the winter. Replant in the spring.

Gypsophila

GYPSOPHILA ELEGANS

Annual
Annual baby's breath, chalk plant

Sowing Directions: Sow seed indoors, 6–8 weeks before the last frost. It is easier, however, to sow outdoors, in midspring when soil can be worked, directly where they are to grow. For continuous summer bloom, try successive sowing every 2–3 weeks for 2–2½ months before the first frost in your area. Just barely cover seed with soil.

Indoor Germination Temperature: 70–80°F

Days to Germination: 10–15

Growing On Temperature: 60–65°F

Garden Planting: Plant after the last frost in full sun (part shade in warmer climates), in well-drained, slightly alkaline soil. Avoid too moist or too acidic soils. Support taller varieties so plants don't lose their bushy shape. Cut back after flowering for a second bloom. Space 10–13 inches apart.

Description: When in bloom, in late spring through fall, these bushy plants, with their light, frothy, small white flowers, create a floating mist of white clouds in the garden. The flowers are used in fresh and dried arrangements. Height: 1–2 feet.

Recommended Cultivar: 'Covent Garden White' produces the largest flowers of the annual baby's breaths. Height: 1½ feet.

GYPSOPHILA

Perennial
G. paniculata: perennial baby's breath (Zones 4–8), *G. repens:* creeping baby's breath (Zones 3–8)

Sowing Directions: Sow indoors 8–9 weeks before the last frost in individual peat pots. Can be sown outdoors, in early spring or fall, directly where they are to grow. Lightly cover with soil.

Indoor Germination Temperature: 70–80°F

Days to Germination: 10–15

Growing On Temperature: 60–65°F

Garden Planting: In the spring, when the soil can be worked, plant in full sun in any well-drained, slightly alkaline soil. Keep well watered during dry spells and cut back after flowering to promote further blooming and to keep a tidy habit. Support taller specimens with stakes or a wire frame. Space 2 feet apart.

Description: Similar to annual baby's breath (see above).

G. paniculata has single or double white flowers that bloom from early to midsummer; longer if cut back after initial flowering. Height: 3 feet.

G. repens is lowgrowing with a creeping habit; it has single or double white, pink, or purple flowers. Height: 6–10 inches.

Recommended Cultivar: 'Early Snowball' is an early-blooming, double white flowering variety. Height: 3 feet.

H

HELENIUM AUTUMNALE

Perennial
Sneezeweed, false sunflower, yellow star
Zones: 3–8

Sowing Directions: Sow indoors 8–10 weeks before the last frost date. Can be sown outdoors, in early spring up to 2 months before the first fall frost, directly where they are to grow. Usually propagated by division. Seeds need light to germinate; lightly press onto surface.

Indoor Germination Temperature: 70–75°F

Days to Germination: 8–12

Growing On Temperature: 55–60°F

Garden Planting: Plant in full sun in any kind of soil, although it thrives in a fertile, slightly acidic soil. Keep well watered during the growing season. Stake the plants as they grow. Deadhead spent flowers. Space 12–14 inches apart.

Description: This tough, native plant produces profuse small yellow, orange, red, and russet brown daisy-like flowers in late summer through fall. An underused perennial, it is capable of adding glorious fall color to the garden and brightness to fresh flower arrangements. Plant in natural drifts or in flowing patterns in beds and borders. Height: 2–4 feet.

HELIANTHEMUM NUMMULARIUM

Perennial
Rockrose, frostweed, sun rose
Zones: 5–9

Sowing Directions: Sow seeds indoors 6–8 weeks before the last frost. Can be sown outdoors, in early spring up to 2 months before the first fall frost, directly where they are to grow Seeds need light to germinate; just barely cover with medium.

Indoor Germination Temperature: 70–75°F

Days to Germination: 15–20

Growing On Temperature: 60–65°F

Garden Planting: Plant in full sun to part shade in any well-drained, sandy loam soil. It is tolerant of poor soils as long as they are alkaline and not soggy. Deadhead flowers as they fade. Cut back the entire plant by one-third or one-half after flowering to maintain an attractive habit and to promote a second bloom period. Space 10–12 inches apart.

Description: This low-growing evergreen native plant has single red, pink, yellow, orange, and light purple flowers that bloom in early summer. Use as a ground cover, in rock gardens, in borders, and as an edging. Height: 8–10 inches.

HELIANTHUS ANNUUS

Annual
Sunflower

Sowing Directions: Sow seeds indoors 2–3 weeks before the last frost into individual peat pots, since plants do not transplant well. For best results, sow seeds outdoors, after the last frost, directly where they are to grow. Space seeds about 12 inches apart for dwarf varieties and about 24–36 inches for larger ones. Plant seeds to a depth of about ¼ inch.

NOTE: *There are several perennial varieties of sunflowers including: H. × multiflorus (Zones 3–10) and H. maximillianii (4–10). They can be grown similarly to annual varieties, but allow more space between plants—about 2–3 feet apart from other plants. Perennials can be cut divided or cut back to the ground in the fall. Perennial varieties will bloom the first year from seed but don't have the large classic annual sunflower heads. They have 3–5 daisylike yellow flowers with smaller dark center. Perennial varieties can be pinched back when young for fuller, bushier plants. They grow 4–9 feet in height.*

Indoor Germination Temperature: 75–85°F

Days to Germination: 10–14

Growing On Temperature: 60–65°F

Garden Planting: After the last frost, plant in full sun in any well-drained soil. Some of the taller varieties may need staking, especially if planted in a windy site. Space 12 inches apart (dwarf); 24–36 inches (large).

TIP:
Mix sunflower cultivars for a special display. For example, try planting 'Sunrise' and 'Sunset' together: Sunrise and sunset are the traditional times of day when heaven and earth seem to meet. With this duo you can experience the dramas of dawn and of dusk during the day throughout the summer.

Description: This stylishly old-fashioned flower is known and loved by everyone. In addition to the traditional bright yellow flowers, newer varieties come in gorgeous shades of bronze, red, mahogany, orange, white, lemon, and tangerine. Height: 15 inches (dwarf), 15 feet or more (giants).

Helianthus annuus 'Sunspot'

Recommended Cultivars:

'Paul Bunyan' hybrid is a giant that seems to bring the magic of folk tales to life right in the garden. Watch the birds flock to this strong-stalked variety when the seeds develop! Height: 15 feet.

'Sunrise' has yellow flowers with chocolatey brown centers. Height: 5 feet.

'Sunset' has glowing flowers of mahogany, red, and gold. Height: 3 feet.

'Sunspot' is a dwarf variety that produces large (12-inch) flowers. Makes a striking container planting, and kids are charmed by this small plant. Height: 18 inches.

HELICHRYSUM BRACTEATUM

Annual
Strawflower, immortelle

Sowing Directions: For best results, sow seeds indoors 4–6 weeks before the last frost. In areas with a long growing season, seeds can be sown outdoors, after the last frost, directly where they are to grow. Seeds need light to germinate; gently press seeds onto surface.

Indoor Germination Temperature: 70–75°F

Helichrysum bracteatum 'Bright Bikinis' mix

Days to Germination: 7–10

Growing On Temperature: 60–65°F

Garden Planting: Plant in full sun in very well-drained soil. Thrives in areas with long, hot growing seasons and will tolerate low-water conditions. Space 10–15 inches apart.

Description: These are the quintessential dried flowers, which, in fact, feel dry while in full bloom on the plant. They come in all manner of glowing colors and color combinations—yellow, red, white, orange, salmon, pink, and purple—and bloom from midsummer to the first frost. Height: 10 inches–4 feet, depending on the variety.

Recommended Cultivar: 'Bright Bikinis' mix: Compact habit with semidouble 2-inch crimson, fiery red, pink, white, and gold flowers. Height: 14 inches.

HELIOPSIS HELIANTHOIDES

Perennial
Oxeye daisy, false sunflower, sunflower heliopsis
Zones: 4–9

Sowing Directions: Sow seed indoors 8–10 weeks before the last frost. Plants started from seed this early will bloom the first year. Can be sown outdoors, in early spring up to 2 months before the first fall frost, directly where they are to grow. Lightly cover seeds with medium.

Indoor Germination Temperature: 65–70°F

Days to Germination: 8–12

Growing On Temperature: 60°F

Garden Planting: Plant in full sun in a rich, loamy, moist, well-drained, slightly acidic soil. If soil is not naturally rich, add some fertilizer early in

the season. Some varieties may need staking. Space 2–2½ feet apart.

Description: Looking look like a cross between a sunflower and a daisy, this plant bears yellow-orange flowers with brown centers from midsummer to early fall. It has an upright habit and makes a long-lasting cut flower. Height: 3 feet.

HELIOTROPIUM ARBORESCENS

Perennial; grow as an annual
Heliotrope
Zones: 9–11

Sowing Directions: Start seed indoors 10–12 weeks before the last frost. It is not recommended to direct sow outdoors. Seeds need light to germinate.

Indoor Germination Temperature: 70°F

Days to Germination: 4–21

Growing On Temperature: 60–65°F

Garden Planting: After the last frost, plant in full sun to part shade, in warmer zones, in rich, well-drained soil. If the soil is poor, fertilize early in the season and again in midsummer. Keep well watered during dry spells. Deadhead to prolong flowering. Space 1½ feet apart.

Description: Known for its delicious fragrance and resplendent blue, violet, and white flowers that cluster in large (6–10-inch) umbel-shaped inflorescences. Use in beds, borders, and containers; plant where it will be easy to enjoy the lovely scent. Height: 8–18 inches.

Recommended Cultivar: 'Dwarf Marine' has large, dark purple flowers and glossy dark foliage. Excellent for container plantings. Height: 8 inches.

HEMEROCALLIS

Perennial
Daylily
Zones: 3–9

Sowing Directions: For indoor sowing, 12 weeks before the last frost, place seeds in a moistened medium in a plastic bag and place in the refrigerator for 6 weeks. After the cold treatment, bring the seeds to 60–70°F for germination. Seeds can be sown outdoors, in late fall or early spring, directly where they are to grow. Usually propagated by division. Cover seeds with about ⅛ inch of medium.

Indoor Germination Temperature: 60–70°F

Days to Germination: 3–50

Growing On Temperature: 55–60°F

Garden Planting: Plant in full sun to part shade in a rich, moist, slightly acidic soil. Will tolerate most well-drained soils. Feed early in the season and provide extra water during

Hemerocallis 'Pompian Rose'

dry periods in the growing season. Cut back flower stems after flowering. Space 2 feet apart.

Description: The workhorse of many summer gardens, this plant adds a casual elegance with its graceful clumps of long, strappy leaves and lovely yellow, orange, red, pink, and lavender lily-like flowers that bloom from early summer on and off to early fall. Mass in borders and use for edgings; use smaller varieties in containers. Excellent naturalizing plant. Height: 2–4 feet.

HESPERIS MATRONALIS

Perennial
Dame's rocket, sweet rocket, garden rocket
Zones: 3–9

Sowing Directions: To get blooms the first year, start seed indoors 8–10 weeks before the last frost date. Can be sown outdoors, in the spring when soil can be worked, directly where they are to grow. Seeds need light to germinate; do not cover with medium.

Indoor Germination Temperature: 75–80°F

Days to Germination: 20–25

Growing On Temperature: 60°F

Garden Planting: Plant in full sun to part shade in any well-drained soil. Remove flower stems after flowering to promote more blooms. Stake taller specimens. Space 12 inches apart.

Description: Produces phloxlike clusters of white to light purple or lilac flowers that bloom from late spring through early summer. The fresh-cut flowers bring fragrance to an arrangement. Use in the middle or back of the border. Height: 4 feet.

HEUCHERA SANGUINEA

Perennial
Coralbells, alumroot
Zones: 3–9

Sowing Directions: Start indoors 8 weeks before the last frost. Seeds can be sown outdoors in fall or early spring. Seeds need light to germinate; gently press onto surface and do not cover with medium.

Indoor Germination Temperature: 60–70°F

Days to Germination: 21–30

Growing On Temperature: 55–60°F

Garden Planting: Plant in full sun to part shade, in hot climates, in a rich, moist, well-drained soil. After flowers fade, cut back stems. It is subject to frost heave, so mulch well over the winter and renew the soil around the base in the spring. Space 14 inches apart.

Heuchera sanguina

Description: A sweet plant with clumps of heart-shaped leaves and delicate white, red, or pink flowers that bloom atop thin stems. Foliage is considered ornamental. Excellent in borders and for edgings. Height: 12–36 inches (in bloom).

HIBISCUS

Annual, perennial
H. coccineus: Hibiscus, scarlet rose mallow (Zones 7–11); *H. moscheutos:* rose mallow, hibiscus, mallow rose, swamp rose (Zones 5–11); *H. acetosella* is an annual.

Sowing Directions: For indoor sowing, scarify seeds with a sharp knife (carefully nick each a half dozen times) or a nail file or piece of sandpaper (rub down the sides of the seeds) and place in warm water to soak overnight before sowing. Start seeds indoors 10–12 weeks before the last frost. For outdoor sowing, sow outdoors, in early spring up to 2 months before first fall frost, directly where they are to grow. Seeds may be scarified and soaked to speed germination. Scatter seeds on the surface and cover very lightly with medium.

Indoor Germination Temperature: 75–80°F

Days to Germination: 7–30

Growing On Temperature: 60–65°F

Garden Planting: Plant in full sun to light shade in rich, moist, well-drained soil. It is important to keep well watered during the growing season if rainfall is low. Since it tolerates moist to wet soils, it can be used as a planting in naturally wet conditions along ponds or in swampy, poorly drained areas. Space 24 inches apart (more for larger varieties).

Hibiscus 'Disco Belle'

Description: One of the more exotic, dramatic, and tropical-looking flowers that grow in North America. Produces large, brilliant, sizzling pink, red, and white flowers from midsummer to early fall. Use as a specimen in a border or grouped to form a seasonal hedge or screen. Height: 2–8 feet.

HOSTA

Perennial
Hosta, plantain lily
Zones: 3–9

Sowing Directions: Sow seeds indoors 6–8 weeks before the last frost. Can be sown outdoors, in early spring up to 2 months before the first fall frost, directly where they are to grow. Plants started from seed will take 2–3 years to bloom. Usually propagated by division. Very lightly cover seeds with medium. Annual hibiscus can be started indoors 6–8 weeks before the last frost, or sown directly outdoors after the last frost.

Indoor Germination Temperature: 70°F

Days to Germination: 15–25

Growing On Temperature: 55°F

Garden Planting: Plant in part to full shade in rich, moist, well-drained soil. Cut off flower stems after flowering. Size varies greatly among varieties. Space 15 inches apart (smaller varieties) or 24–36 inches apart (larger varieties).

NOTE: *Remember that most hosta cultivars will not come true from seed.*

Description: Very popular shade plant that is available in hundreds of types in all shades of green and blue with many variegated varieties of green and white, yellow and green, and gold and green. White or lavender flowers are borne on stalks above the foliage and bloom in the summer. Use for borders, massed as a ground cover, or as specimen plants for contrast. Susceptible to slugs, which love to chew the leaves. Height: 1–3 feet.

HYPOESTES PHYLLOSTACHYA

Annual
Polka-dot plant

Sowing Directions: For outdoor planting, it is best to start seeds indoors 4–6 weeks before the last frost. Can be sown outdoors, after the last frost, directly where they are to grow. For a houseplant, sow seed indoors anytime of the year.

Indoor Germination Temperature: 70–75°F

Days to Germination: 9–14

Growing On Temperature: 60–65°F

Garden Planting: Plant in part shade in any well-drained soil. In cooler northern gardens, will tolerate full sun, especially if shaded by nearby taller plants. To increase bushiness, pinch growing shoot tips off early in the growing season. Space 12 inches apart.

Description: Grown for its foliage, which looks as if pink, red, or white paint had been splashed on it. Produces meager light purple flowers. Use for edgings, borders, and mixed container planting. Can be massed as a bedding plant. Height: 5–18 inches.

Recommended Cultivar: 'Splash' mix plants produce green foliage splashed with white, rose, or pink. Height: 5 inches.

I

IBERIS

Annual, perennial
I. amara, I. umbellata: (annual)
Hyacinth-flowered candytuft, globe candytuft; *I. sempervirens*: (perennial)
Candytuft, evergreen candytuft
Zones: 3–8

TIP: Polka-dot plants can be dug up at the end of the growing season before the first frost and brought inside as a houseplant. Kids like growing this plant for its funny spotted leaves.

TIP:

In northern climates, cover candytuft foliage with evergreen boughs to protect it against drying damage in the winter.

Sowing Directions: For both annual and perennial candytuft sow seeds indoors 6–8 weeks before the last frost. Can be sown outdoors, in early spring, where it is to grow. Plants grown from seed will flower the second year. Usually propagated by division. Cover lightly with about ¼ inch of medium.

Indoor Germination Temperature: 60–65°F

Days to Germination: 10–21

Growing On Temperature: 55–60°F

Garden Planting: Plant in full sun to part shade in rich, well-drained soil. Cut plants back after flowering. Space 12 inches apart.

Description: Perennial candytuft is a low-growing, spreading plant with delicate, tiny, densely clustered white flowers that bloom in mid-spring. Its dark green foliage is evergreen throughout the year. Use in edgings, borders, and rock gardens, and—best of all—for spilling over stone walls or other structures. Height: 6–10 inches. Globe and hyacinth-flowered candytuft are upright, bushy annuals with white, purple, pink, or red flowers. Height: 8–12 inches.

ICE PLANT. SEE MESEMBRYANTFHEMUM CHRYSTALLINUM (SYN. DOROTHEANTHUS)

IMPATIENS

Annual
I. wallerana: impatiens, busy Lizzy, balsam, touch-me-not

Sowing Directions: Sow seeds indoors 8–10 weeks before the last frost. For best results, provide a uniform potting mix temperature and high humidity. Mist seeds daily and cover the container with glass or clear plastic. Since these seedlings are susceptible to damping off, make sure they are not overwatered and use very clean containers. Seedlings grown in vermiculite only will be less prone to damping off than if grown in other soilless germination mixes. Once the seeds have germinated, prop up the cover to allow air circulation. Seeds need light to germinate; do not cover with medium.

Indoor Germination Temperature: 70–75°F

Days to Germination: 10–18

Growing On Temperature: 60°F

Garden Planting: Plant in part to full shade in rich, moist, well-drained, slightly acidic soil. Fertilize early in the growing season and keep it well watered throughout the growing season. If planted in a container, make sure it receives adequate water. Space 12 inches apart.

Description: As one of the nation's most popular summertime flowers, it brightens up shady places all over America. Easy and reliable for unbeatable color throughout the season, plants are available in red, pink, white, orange, purple, and even yellow ('African Queen'). New Guinea hybrid impatiens have gorgeous variegated tropical-looking foliage with brightly colored flowers. Double hybrid impatiens are unusually beautiful and lush and look like shade-blooming roses. Use massed in beds and borders. Height: 8–24 inches.

Recommended Cultivars:

'Dazzler' series and 'Super Elfin' series: Both types produce masses of blooms on plants that do not need pinching for good branching and shape.

Impatiens 'Rosette' mix

'Rosette' mix produces full-double, semidouble, and single flowers of brilliant red, pink, salmon, blush, white, and bicolored red-and-white flowers. Height: 18–20 inches.

IPOMOEA

Annual
I. alba: moonflower, *I. tricolor:* morning glory

Sowing Directions: For best results, scarify the seeds by nicking with a sharp knife or sanding a bit with a nail file or sandpaper. Then soak the scarified seeds in warm water for 24 hours. Sow treated seeds indoors 4–6 weeks before the last frost. Sow seeds in individual peat pots (about two seeds per pot) to avoid transplant shock. Or sow treated seeds outdoors, after the last frost, directly where they are to grow.

Indoor Germination Temperature: 70–80°F

Days to Germination: 5–7

Growing On Temperature: 60–65°F

Garden Planting: Plant in full sun in any well-drained soil. More profuse blooms will be produced if planted in low-fertility soil. In low rainfall periods, provide *I. alba* with extra water; *I. tricolor* is more drought-tolerant.

Description: Plant these fast-growing vines near a trellis, fence, wall, or other support or try growing them up through a shrub.

Ipomoea 'Morning Glory'

I. alba has a vigorous vine with large, luminous white fragrant flowers that bloom in the evening throughout the summer and into early fall. Height: 15 feet in a season.

I. tricolor is the classic old-fashioned blue, pink, and white

TIP:
Keep on eye on your morning glory, as it can be invasive.

vine with flowers that bloom in the early morning and close by early afternoon. Height: 8–10 feet in a season.

Recommended Cultivars:

I. alba 'Giant White' produces 6-inch white flowers. Height: 15 feet.

I. tricolor 'Heavenly Blue' produces soothing 5-inch sky blue flowers with white throats. Height: 8 feet.

IRIS

Perennial
Iris

NOTE: *It is very difficult to start iris seeds. The beginner is encouraged to buy plants, bulbs, or rhizomes for later propagation by division. However, general germinating instructions that cover the most commonly grown iris species are presented here.*

Sowing Directions: Seeds are sown outdoors in the fall or in the spring. Soak seeds in warm water for at least 24 hours before planting. For fall sowing, sow seeds into containers that are placed in the ground in a sheltered area for the winter. Bring the containers inside in the early spring and germinate at 70–75°F. For spring sowing, sow seeds in a container and cover it with a plastic bag. Place it in a refrigerator for 4–6 weeks before bringing the container out to germinate at 70–75°F.

Germination time varies with the species, but it usually takes 1–2 months; though germination can be slow and inconsistent and can take up to 1½ years.

Whether fall or spring sown, after germination, thin seedlings so each has a 6-inch area within a large container or so that there is one seedling per 6-inch container. Grow seedlings two full seasons in their containers in a sheltered, part shade site in the garden before transplanting them to their permanent growing space.

KALE, FLOWERING. SEE BRASSICA OLERACEA ACEPHALA GROUP

KNIPHOFIA UVARIA

Perennial
Red-hot poker, torch lily, tritoma
Zones: 5–9

Sowing Directions: Start seed indoors 6–8 weeks before planting out in the spring, 4–6 weeks before the last frost. Can be sown outdoors, in early spring and up to 2 months before the first fall frost, directly where they are to grow. Plants grown from seed may bloom in the second year, but peak bloom will be in the third. Usually propagated by division. Cover seed lightly by about ¼ inch.

Indoor Germination Temperature: 70–75°F

Days to Germination: 21–28

Growing On Temperature: 60°F

Garden Planting: Plant in full sun in any well-drained soil. It will not tolerate heavily wet soils. For best results, avoid transplanting; it likes to be left alone. Cut back flower stalks after flowering. Space 18 inches apart.

Kniphofia Uvaria

Description: An unusually stunning and dramatic perennial that produces fiery red, orange, coral, cream, white, and yellow torchlike raceme flowers in the summer. It is very easy to care for, thriving on neglect in the proper growing conditions. Use in a border. Height: 3–4 feet.

KOCHIA SCOPARIA FORMA TRICHOPHYLLA

Annual
Summer cypress, burning bush, belvedere, fire bush

Sowing Directions: Germination rates greatly improve if the seed is soaked overnight in warm water before planting. Sow seeds indoors 4–6 weeks before the last frost. Can be sown outdoors, after the last frost, where they are to grow. Always use fresh seed, as older seed will not be viable. Seeds need light to germinate; press lightly onto the surface and do not cover with medium.

Indoor Germination Temperature: 70–75°F

Days to Germination: 10–15

Growing On Temperature: 65°F

Garden Planting: Plant in full sun in well-drained, dryish soil. Provide a little extra fertilizer early in the growing season and then again about midsummer to keep the plant a brilliant green color. Space 24 inches apart.

Description: This bushy, round, light green foliage plant with its finely cut, ornamental leaves is a big attraction since it looks like a little evergreen shrub. In the fall, the foliage turns an intense, glowing red. Makes a nice contrast to brightly colored red- and yellow-flowering plants in the summer. Use for mini hedges and screens, for edgings, or as an accent plant in the border. Height: 2–3 feet.

L

LANTANA CAMARA

Perennial; grown as an annual
Lantana
Zones: 9–11

Sowing Directions: Soak seeds overnight in warm water before germinating. Sow seeds indoors 8–10

weeks before the last frost. For Zones 9–11, can be sown outdoors, in early spring, where they are to grow. Cover seeds with medium to a depth of ⅛ inch.

Indoor Germination Temperature: 70–75°F

Days to Germination: 30–55

Growing On Temperature: 60–65°F

Garden Planting: Plant in full sun to light shade in well-drained soil. Keep well watered throughout the growing season. Pinch back for a bushier habit. Space 18–24 inches apart.

Description: A gorgeous, heat-loving plant with yellow, orange, white, and red clusters of small flowers that bloom from midsummer to frost. Many of the verbenalike flowers have a bicolor effect. Use as specimen plants in the border or in container plantings. Susceptible to whitefly. Height: 1–2 feet.

> **NOTE:** *If you like butterflies in your garden, plant lantana. This plant emits an irresistible siren song to a range of butterflies and hummingbirds.*

LARKSPUR. SEE *CONSOLIDA AMBIGUA*

LATHYRUS ODORATUS

Annual
Annual sweet pea, sweet pea

Sowing Directions: Before sowing (either indoors or outdoors), nick the seeds with a sharp knife or sand

down the coat a bit with sandpaper or a nail file; then soak the scarified seeds in warm water for 24 hours. For indoor sowing, 4–6 weeks before the last frost sow one to two treated seeds into individual peat pots, to minimize transplant shock. Cover seeds with 1 inch of potting mix. For outdoor sowing, in early spring sow treated seeds directly where they are to grow. Cover seeds with 2 inches of soil. Seeds need darkness to germinate.

> **NOTE:** *In Zones 8–11, seeds can be sown outdoors, in late fall and early winter, directly where they are to grow for winter and early spring flowering.*

Indoor Germination Temperature: 55–65°F

Days to Germination: 10–15

Growing On Temperature: 50–55°F

Garden Planting: After the last frost, plant in full sun in well-prepared, organically rich, fertile, well-drained soil. Keep the soil evenly moist. Provide some kind of trellis or frame support for vining varieties. Space 8–12 inches apart.

Description: There are few flowers with the immediate charm and old-fashioned appeal of these with their gentle, bright colors and sweet fragrance. The pealike flowers in pink, white, blue, red, purple, and rose bloom in spring through early summer. This cool-weather plant is available in either a bush or a vine habit. In warm areas, choose heat-tolerant cultivars. Height: 15–24 inches (bush), 10 feet (vine).

TIP:

Lantana can be dug up in the fall before the first frost. Pot it up and grow it indoors as a houseplant for the winter. It can be planted outdoors again the following spring, if desired. It can also be trained into a standard.

Recommended Cultivars:

'Pink Perfume': A dwarf, mounding plant, with deliciously fragrant shell pink flowers that are especially nice for containers. Height: 15 inches.

'Sweet Dreams': A heat-tolerant newer hybrid with large, gently ruffled, lavender, pink, and white flowers. Height: 6–8 feet.

NOTE: A perennial variety of sweet pea, L. latifulius climbs 4–9 feet in height with clusters of red, white, pink, blue, and purple flowers. It can be started and grown similarly to the annual variety. Germination takes a bit longer—14–21 days.

LAVATERA

Annual
Annual mallow, tree mallow, mallow

Sowing Directions: Start seeds indoors 6–8 weeks before the last frost. Sow seeds into individual peat pots that can be directly transplanted into the garden. Seeds can be sown outdoors, in early to midspring, where they are to grow. For continuous summer blooms, sow successively every two weeks until early summer. Lightly cover seeds with soil.

Indoor Germination Temperature: 70°F

Days to Germination: 15–20

Growing On Temperature: 60–65°F

Garden Planting: Plant in full sun in any well-drained soil. Plants thrive

with some extra fertilizer regularly applied throughout the growing season. Deadhead to promote more flowering. Stake taller growing varieties. Space 24 inches apart.

Lavatera 'Silver Cup'

Description: With a shrublike habit, this plant produces delectable hollyhock-shaped neon pink, purple, and white flowers that bloom throughout the summer. Use in borders or as an annual hedge or screen. Height: 2–4 feet.

LAVENDER. SEE *LAVANDULA* (IN CHAPTER 6)

LAYIA PLATYGLOSSA

Annual
Tidytips

Sowing Directions: Sow seeds indoors 6–8 weeks before the last frost Can be sown outdoors, after the last frost, directly where they are to grow. Cover seeds with medium by about ⅛ inch.

Indoor Germination Temperature: 70–75°F

Days to Germination: 8–12

Growing On Temperature: 60–65°F

Garden Planting: Plant in full sun in a light, well-drained soil. Keep well watered; if the soil is poor, fertilize monthly throughout the growing season. Prefers cooler growing temperatures. Space 15 inches apart.

Description: A fast-growing annual with a mounding habit that produces bright yellow, white-tipped, daisy-like flowers throughout the summer. Height: 12–24 inches.

LEWISIA REDIVIVA

Perennial
Bitterroot
Zones: 4–8

Sowing Directions: For sowing indoors, place seeds in a container or plastic bag with moistened medium and refrigerate or freeze for 1 month. Sow treated seeds in containers; do not cover the seed. Allow seedlings to grow for 1 year in the container before transplanting them to their permanent site in the garden. For best results, sow outdoors, in late fall, directly where they are to grow. Scatter seeds onto the soil surface and press lightly.

Indoor Germination Temperature: 60–65°F

Days to Germination 1–90 or longer

Growing On Temperature: 55–60°F

Garden Planting: Plant in full sun to part shade in sandy, very well-drained soil. Make sure soil is quite well drained, as this plant is very sensitive to rot problems. Space 6–8 inches apart.

Description: This low-growing, native to the California mountains has succulentlike narrow leaves and bears star-shaped pink or white flowers that bloom in spring through early summer. Use in rock gardens and other very well-drained, sandy or gritty sites in a border. Height: 4–6 inches.

LIATRIS SPICATA

Perennial
Gayfeather, blazing star, prairie snakeroot
Zones: 3–8

Sowing Directions: Sow seeds indoors 6–8 weeks before the last frost. For best results, sow seeds outdoors, in early spring and up to 2 months before the first fall frost, directly where they will grow. Cover seeds with medium by ⅛ inch.

Liatris

Indoor Germination Temperature: 65–70°F

Days to Germination: 21–28

Growing On Temperature: 50°F

Garden Planting: Plant in full sun in rich, well-drained, loamy soil. It will tolerate in poorer soils as long as there is good drainage. Space 15 inches apart.

Description: This stately, heat- and drought-tolerant, erect-growing native plant explodes with pink and purple woolly flowers all along the upper stem throughout the summer to early fall. Lower stems bear bits of grasslike spiky foliage. Use in the back of the border or in naturalized plantings; makes a long-lasting cut flower. Height: 3–6 feet.

LIMONIUM SINUATUM

Annual
Sea lavender, statice

Sowing Directions: For best results, sow these seeds indoors 8–10 weeks before the last frost. In more southern areas with long growing seasons, they can be sown outdoors, after the last frost, where they are to grow. Lightly cover seeds with medium by ⅛ inch.

Indoor Germination Temperature: 65–75°F

Days to Germination: 14–21

Growing On Temperature: 50–55°F

Garden Planting: Plant in full sun in light, well-drained soil. Use this drought-tolerant plant for hot climates and dry soils. Space 15 inches apart.

Description: This unique plant is grown and prized for its dry, papery flower sprays of vivid yellow, white, pink, purple, blue, apricot, and red that bloom in the summer on very stiff stems. The cut dried flowers last for months, if not years in arrangements and crafts. Use in beds, borders, and rock gardens to attract butterflies. Height: 12–34 inches.

Limonium sinuatum 'Art Shades' mix

Recommended Cultivar: 'Art Shades' mix produces understated pink, mauve, apricot, and yellow flowers. Height: 30 inches.

LINUM PERENNE

Perennial
Perennial flax
Zones: 5–9

Sowing Directions: Sow seeds indoors 6–8 weeks before the last frost into individual peat pots that can be directly transplanted to their permanent garden sites, to avoid transplant shock. For best results, sow outdoors, in early spring and up to 2 months before the first fall frost, directly where they are to grow. Lightly cover seeds with medium by ⅛ inch.

Indoor Germination Temperature: 65–75°F

Days to Germination: 20–25

Growing On Temperature: 55–60°F

TIP:

If you want your lilyturf to produce berries, do not cut it back after it blooms.

Garden Planting: After the last frost, plant in full sun in any soil with good drainage. It prefers slightly alkaline conditions. Cut back after flowering to promote more blooms. Space 12–18 inches apart.

Description: One of the most beautiful of the all-blue flowering plants, it produces sky blue, satiny flowers from early to late summer. Use in clusters in borders and rock gardens. Height: 1½–2 feet.

LIRIOPE MUSCARI

Perennial
Lilyturf
Zones: 6–9

Sowing Directions: Before germinating, soak seeds in warm water for 24 hours. Sow seeds indoors 6–8 weeks before the last frost date. Can be sown outdoors in early spring and up to 2 months before the first fall frost. Cover seeds with medium by about ¼ inch.

Indoor Germination Temperature: 65–70°F

Days to Germination: 25–30

Growing On Temperature: 55–60°F

Garden Planting: After the last frost, plant in part or full shade in moist, rich, well-drained soil. It will tolerate dry soils in the shade. To plant in full sun, the soil must be kept consistently moist. Keep well watered if rainfall is low during the summer. Space 18–24 inches apart.

Description: A very low-maintenance plant with a clumping habit and grasslike foliage in dark green or variegated in yellow and green.

Purple or white flowers are borne up, over, and through the foliage on spikes from late summer through fall. After flowering, attractive black berries form on the spikes. Excellent for use as a ground cover or edging plant. Height: 1½ feet.

Lobelia

LOBELIA ERINUS

Annual
Annual lobelia

Sowing Directions: Start seeds indoors 10–12 weeks before the last frost. This seed is extremely fine and small. It is easiest to sow it with a little bit of sand mixed into the packet to get even distribution over the container surface or to sow seed in rows. When thinning out seed, seedlings are usually pricked out in little patches to transplant and grow on, since it is hard to separate the seedlings without injury. Sow seed on the surface of the medium and lightly press into the soil. Seeds need light to germinate; do not cover with medium.

Indoor Germination Temperature: 70–80°F

NOTE: *Lobelia is susceptible to damping off so make sure containers are thoroughly washed. A sterile, soilless mix or vermiculite alone should be used. Water plants from below and give the roots proper air circulation after the seedlings emerge.*

Days to Germination: 14–21

Growing On Temperature: 60°F

Garden Planting: After the last frost, plant in full sun or part shade in a rich, well-drained soil. It prefers cool temperatures, so plant in part shade in warmer areas. When planted in full sun, it will be stressed and flowering will not be as profuse. Keep it well watered throughout the summer, especially during dry spells. Space 6–8 inches apart.

Description: This plant provides nonstop abundant billows of tiny purple, white, or red flowers all summer and into the fall. Both low-growing upright and trailing varieties are available. Use as a bedding, rock garden, or edging plant. Makes a gorgeous container plant, particularly the trailing varieties. Height: 4–6 inches (upright), several feet (trailing).

Recommended Cultivar: 'Crystal Palace' produces intense cobalt blue flowers with unusual and handsome bronzy foliage. Height: 6 inches.

LOBELIA CARDINALIS

Perennial
Cardinal flower, perennial lobelia
Zones: 2–9

Sowing Directions: To sow indoors, seeds must first be given a cold treatment. Place seeds in moistened medium in a container and cover with a plastic bag; then refrigerate for 10–12 weeks before taking out to germinate. It is easier to sow seeds outdoors, in the fall, where they are to grow. Mature plants will often self-sow. Seeds need light to germinate; do not cover seeds with medium.

Indoor Germination Temperature: 70–75°F

Days to Germination: 14–20

Growing On Temperature: 60–65°F

Garden Planting: Plant in full sun to part shade (although it prefers full sun) in rich, moist, well-drained soil. Keep it well watered if rainfall is low during the summer. Space 18 inches apart.

Description: This strong, stately, and vibrant plant has ravishing red flowers that bloom along erect stems for several weeks starting in late summer. Use in borders and naturalized plantings. Height: 2–4 feet.

Lobelia Cardinalis

L. siphilitica: Blue-flowering perennial lobelia is germinated and grown similarly to *L. Cardinalis*. It blooms from mid to late summer into fall. Height: 2–3 feet.

LOBULARIA MARITIMA

Annual
Sweet alyssum

Sowing Directions: Start seeds indoors 4–6 weeks before the last frost. Can be sown outdoors, in early spring after danger of heavy frost, where it is to grow. Seeds are very fine so mix with a little sand for more even distribution. Mature plants often self-sow. Seeds need light to germinate; gently press seed onto the surface and do not cover with medium.

Indoor Germination Temperature: 75–80°F

Days to Germination: 8–10

Growing On Temperature: 60–65

NOTE: *Alyssum seedlings can be difficult to separate. It's all right if they are overcrowded.*

Garden Planting: Plant in full sun to light shade in well-drained soil. In hot climates, it will do best in light shade. Try to keep the soil consistently moist and provide extra water during dry spells. Shear back after flowering to stimulate more growth and flowers throughout the season. Late in the summer, growth may stall because of the heat; it will revive with the onset of cooler temperatures. Space 8–10 inches apart.

Description: A dependable summer performer with masses of tiny fragrant flower clusters in shades of white, pink, lavender, and purple that bloom most of the summer. Use in borders, rock gardens, and edgings or allow it to tumble out of a hanging basket or cascade down the side of a container. Height: 3–6 inches.

Recommended Cultivar: 'New Carpet of Snow' produces mounds of snow white flowers that bloom all summer on plants that spread to 8 inches wide. Height: 4 inches.

LUNARIA ANNUA

Biennial
Money plant, honesty, silver dollar, moneywort, dollar plant
Zones: 3–8

Sowing Directions: Sow seeds indoors 8–10 weeks before the last frost. For best results, sow seeds outdoors, in early spring, directly where they are to grow. Plants grown from seed will bloom the second year. Lightly cover seed with medium by ⅛ inch.

Indoor Germination Temperature: 65–75°F

Days to Germination: 10–14

Growing On Temperature: 55–60°F

Garden Planting: After the last frost, plant in full sun to light shade in well-drained soil.

Description: Grown for its unusual,

Lunaria annua

round, silvery white, transparent seedpods that develop in midsummer, this upright, rounded plant produces white or purple flowers that bloom in late spring. As its name implies, the seedpods resemble silver dollars or small coin purses. When dried, the pods are used in fresh and dried flower arrangements. Use in naturalized plantings, and watch carefully as it can become invasive. Height: 24–30 inches.

LUPINUS

Perennial
Lupine, perennial lupine
Zones: 4–8

Sowing Directions: Before sowing seeds, scarify them with a sharp knife, sandpaper, or a file; then soak them in warm water for 24 hours before sowing. Sow indoors 6–8 weeks before the last frost into individual peat pots that can be directly transplanted into the garden, to avoid transplant shock. For best results, sow seeds outdoors, in early spring, where they are to grow. Lightly cover the seeds. Blooms appear early to midsummer (June–July).

Indoor Germination Temperature: 80°F (daytime), 70°F (nighttime)

Days to Germination: 25–30

Growing On Temperature: 55–60°F

Garden Planting: Plant in full sun to light shade in moist, well-drained, slightly acidic soil. Mulch and keep well watered, especially during dry periods. Cut back plants after flowering. Resents being transplanted so once established in the garden, leave it alone. Does best in cooler climates and will not thrive under hot growing conditions. Space 18–24 inches apart.

Lupinus 'Russell' hybrids

Description: A glorious plant with stunning form and color. The upright stems bear intensely colored blue, yellow, pink, purple, white, and bicolor pealike flowers that cluster and bloom along the top of the stems. Plant in drifts in the back of the border and massed throughout naturalized garden areas. Height: 2–5 feet.

NOTE: *L. subcarnosus (bluebonnet, Texas bluebonnet) is an annual with similar germinating and growing conditions as L. hybrids. Bluebonnet germinates in 15–20 days. It blooms in spring and early summer. Height: 1 foot.*

TIP:
Plant lupines where they can be seen up close. When in bloom, these plants are architecturally fascinating in their structural formation. A massed display of lupines in peak bloom is unforgettable.

LYCHNIS CHALCEDONICA

Perennial
Maltese cross, campion, Jerusalem cross
Zones: 3–9

Sowing Directions: Sow seed indoors 10–12 weeks before the last frost. Can be sown outdoors in early spring and up to 2 months before the first fall frost. Plants grown from seed will bloom the first year, if started early. Seeds need light to germinate; gently press seeds onto the surface; do not cover with medium.

Indoor Germination Temperature: 70°F

Days to Germination: 21–25

Growing On Temperature: 60–65°F

Garden Planting: Plant in full sun to light shade in moist, slightly acidic, fertile, very well-drained soil. It will grow in dry soils, but much prefers moist conditions. Cutting back plants after flowering may result in a second bloom period. Space 12–15 inches apart.

> **NOTE:** *L. Coronaria (rose campion) is germinated and grown as L. chalcedonica. Rose campion has gorgeous and intensely colored magenta-rose, pink, or white flowers that bloom from late spring to early summer. The foliage is silvery gray. Height: 2–3 feet.*

Description: These regal, brilliantly rich, scarlet cross-shaped tiny flowers cluster to form a dome at the top of tall erect stems in the summer. Plant in groups in the border and in drifts in naturalized garden areas. Height: 3 feet.

LYSIMACHIA CLETHROIDES

Perennial
Gooseneck loosestrife, Japanese loosestrife
Zones: 3–8

Sowing Directions: For best results, sow seeds outdoors directly in the garden in the fall or early spring to midsummer. Sow in a protected area and transplant to a permanent site when seedlings reach 4–6 inches tall. Usually propagated by division.

Germination Temperature: 65–70°F

Days to Germination: 30–90

Growing On Temperature: 55–60°F

Garden Planting: Plant in full sun to part shade in any soil, although it prefers moist conditions. Cut back plants in the fall. Space 18 inches apart.

Description: One can almost hear a cacophony of honking coming from a massed planting in full bloom. The long, bent white flowers look like a group of botanical geese that have landed in the garden. Its deep dark green foliage turns an attractive bronzy red in late summer through fall. Height: 1–3 feet.

LYTHRUM SALICARIA

Perennial
Purple loosestrife, loosestrife, red Sally
Zones: 4–8

TIP: Gooseneck loosestrife is a vigorous grower and will become invasive if not thinned out regularly throughout the growing season.

Sowing Directions: Sow seed indoors 6–8 weeks before the last frost. Can be sown outdoors, in early spring or late fall, where they are to grow. Lightly cover seeds.

Indoor Germination Temperature: 65–70°F

Days to Germination: 14–30

Growing On Temperature: 55–60°F

Garden Planting: Plant in full sun to light shade in slightly acidic, well-drained soil. It will grow in dry or moist soils and tolerates wet soils

NOTE: *Purple loosestrife becomes dangerously invasive in wetland areas. Some states have banned it from being planted, because it chokes out other plants, upsetting the ecological balance in natural areas. Watch this plant carefully in your garden.*

Lythrum Salicaria

very well—in fact, too well. Space 24 inches apart.

Description: A very easy-to-grow perennial with lovely spikes of pink and reddish purple flowers that bloom densely in mid to late summer along upright growing stems. Height: 3–4 feet.

Recommended Cultivars: These two cultivars are good choices for the home garden and are the least invasive of the many available cultivars.

'Happy' has dark pink flowers and is low growing. Height: 15–18 inches.

'Marden's Pink' has pink flowers. Height: 3 feet or more.

M

MACHAERANTHERA TANACETIFOLIA

Annual
Tahoka daisy

Sowing Directions: For best results, give seeds a cold treatment by placing them in a refrigerator for 2 weeks before sowing. Sow seeds indoors 6–8 weeks before the last frost. Can be sown outdoors, in early to midspring, directly where they are to grow.

Indoor Germination Temperature: 70–75°F

Days to Germination: 21–30

Growing On Temperature: 60°F

Garden Planting: After the last frost plant in full sun to light shade in any well-drained soil. Does best in cooler areas. Space 12 inches apart.

Description: This blue and light lavender daisy flower has a cheerful yellow center and blooms all summer to frost. Use it in the border or as a cut flower. Height: 18–24 inches.

MACLEAYA CORDATA

Perennial
Plume poppy, tree celandine
Zones: 3–9

Sowing Directions: For best results, sow seed directly outdoors in early spring up to 2 months before the first fall frost. Established plants will readily self-sow. Usually propagated by division. Very lightly cover the seeds with soil.

Germination Temperature: 60–65°F

Days to Germination: 10–14

Growing On Temperature: 55–60°F

Garden Planting: Plant in full sun to light shade in any moist, well-drained soil. Space at least 3–4 feet apart.

Description: This rampant and vigorous plant is not for the faint of heart. Yet the downy, creamy white, airy flower plumes that bloom atop large, deeply lobed, figlike leaves in mid to late summer are very beautiful. Height: 8–10 feet in one season.

MALTESE CROSS. SEE LYCHNIS CHALCEDONICA

MALVA ALCEA VAR. FASTIGIATA

Perennial
Mallow, hollyhock mallow
Zones: 4–9

Sowing Directions: Sow seed indoors 6–8 weeks before the last frost. Can be sown outdoors in spring to midsummer. Established plants will self-sow. Very lightly cover seeds with medium.

Indoor Germination Temperature: 70–75°F

Days to Germination: 5–14

Growing On Temperature: 60°F

Garden Planting: Plant in full sun or very light shade, in warmer climates, in any well-drained soil. It will tolerate dry soils but will do best if given extra water during dry spells. Apply fertilizer in the spring. Cut back plants in late fall. Stake taller specimens. Space 24 inches apart.

Description: An upright growing plant with a loose shrubby habit and lovely white, pink, and light lavender hollyhocklike flowers that bloom from midsummer through to fall. Use in borders and for achieving a cottage garden effect. Height: 4–6 feet.

NOTE: *M. moschata (musk mallow) is similar to M. Alcea but grows to just 2–2½ feet.*

MARIGOLD. SEE *TAGETES*

MATTHIOLA INCANA

Annual
Stock, common stock, ten week
stock, gillyflower

Sowing Directions: Sow indoors
4–6 weeks before the last frost. Sow
outdoors after the last frost. Seeds
need light to germinate; lightly press
seeds onto surface; do not cover
with medium.

Indoor Germination Temperature:
55–60°F

Days to Germination: 7–14

Growing On Temperature: 50–55°F

Garden Planting: Plant in full sun or
light shade, in warmer climates, in
rich, moist, well-drained soil. It
prefers cooler growing conditions
for best flowering and growth. Keep
it well watered and provide regular
fertilizer feedings throughout the
growing season, if the soil is not naturally fertile. Space 12 inches apart.

Description: This bushy plant produces soft white, red, purple, pink,
and blue single and double flowers
that bloom along erect spikes in
early to midsummer. Plant close to a
patio or seating areas to enjoy the
delightful fresh fragrance. Use in the
cutting garden or border. Height:
1–2½ feet.

Recommended Cultivar: 'Giant
Imperial' mix has wonderfully fragrant white, pink, and lavender
flowers. Height: 1½ feet.

MECONOPSIS GRANDIS

Perennial
Blue poppy, meconopsis
Zones: 6–8

Sowing Directions: Sow seeds
indoors 6–8 weeks before the last
frost. Seeds can be difficult to germinate and seedlings are known for
having damping off problems. Sow
seeds in clean containers using a sterile growing medium. Lightly press
seeds onto the surface as seeds need
light for best germination. Use a
sifter and lightly cover seeds with a
bit of the potting medium. Be careful
not to overwater seedlings; it's best
to water from the bottom. Give
seedlings good air circulation. Prick
out seedlings and transplant them to
a larger container when they have
developed two true leaves.

NOTE: *Some gardeners recommend pinching off the
flower buds for the first season to encourage more foliage
growth to make it a strong,
long-lived plant.*

TIP:
Don't wait too
long in the spring
to sow stock
seeds, since hot
weather will stall
flowering.

Meconopsis grandis

Indoor Germination Temperature: 55–65°F

Days to Germination: 21–30

Growing On Temperature: 55–60°F

Garden Planting: Plant in part to full shade in moist, rich, acidic, well-drained soil. It prefers cool summer nights; make sure it has extra water during hot, dry periods. Although it tends to be short lived, it is exquisite. Cut plants back in the fall. Space 2 feet apart.

Description: A sought-after and prized plant, it has large (4–5-inch) sky blue poppylike flowers with gold or light yellow centers that bloom in early summer. The flowers are truly one of the most gorgeous, other-worldly shades of silky blue. Height: 3–5 feet.

MERTENSIA VIRGINICA

Perennial
Virginia bluebells, cowslip, bluebells, mountain bluebells
Zones: 3–8

Sowing Directions: For best results, sow seeds outdoors, in early spring through early summer, where they are to grow or into containers that are placed in a protected garden area. Seedlings grown in containers can be transplanted to a garden site in the fall or in the spring of the following growing season. Plants grown from seed will usually take 3 years to bloom. Established

> **NOTE:** *It is best to use fresh seed, which is available in early to midsummer.*

plants will self-sow. Usually sold as roots for division. Lightly cover seeds with soil.

Germination Temperature: 60–65°F

Days to Germination: 30–45

Growing On Temperature: 50–55°F

Garden Planting: Plant in part to full shade in moist, rich, humusy soil. It needs a good mulch covering to keep the soil cool and moist during the summer. Space 12 inches apart.

Description: One of the eagerly anticipated rites of early spring is to see this plant in full bloom with its clusters of small purple and blue trumpet-shaped flowers. Displays attractive bluish green leaves. Use in shady borders, beds, and woodland areas. Height: 1–2 feet.

MESEMBRYANTHEMUM CHRYSTALLINUM (SYN. DOROTHEANTHUS)

Annual
Ice plant, Livingstone daisy

Sowing Directions: Sow seeds indoors 10–12 weeks before the last frost. Can be sown outdoors, after the last frost, directly where they are to grow. Press seeds onto the surface, as they are quite fine. Seeds need darkness to germinate; if sowing indoors, cover the container with several layers of newspaper.

> **NOTE:** *For continuous bloom throughout the growing season, sow seeds in 2-week intervals outdoors from early spring through early summer.*

TIP:
Virginia bluebells die back in the summer. You'll want to fill in any conspicuous gaps with annuals.

Indoor Germination Temperature: 65–75°F

Days to Germination: 15–20

NOTE: *Ice plant is suscepti-ble to soil fungal problems. Thus it is best to grow this plant in a different location in the garden each year.*

Growing On Temperature: 60°F

Garden Planting: Plant in full sun in dry, sandy, well-drained soil. Space 8–12 inches apart.

Description: Bright yellow, white, pink, and red daisylike flowers bloom in spring and summer. A good ground cover for hot, dry climates, as it's very drought- and heat-tolerant and spreads out almost twice as long as it is high. Use as bedding, for edging, or for container plantings. Height: 6–12 inches.

Mesembryanthemum chrystallinum 'Lunette'

Recommended Cultivar: 'Lunette' is an excellent annual ground cover for hot, dry areas with sandy soils. It produces yellow flowers with bright red centers. Height: 6 inches.

MIMULUS × HYBRIDUS

Annual
Monkey flower

Sowing Directions: For best results, refrigerate seeds for 4 weeks before germinating. Sow seeds indoors 10–12 weeks before the last frost. Seeds need light to germinate; gently press seeds onto surface; do not cover.

Indoor Germination Temperature: 60–70°F

Days to Germination: 5–14

Growing On Temperature: 55–60°F

Garden Planting: Plant in full sun to light shade in rich, well-drained, humusy, slightly acidic soil. It prefers cool growing conditions and should be planted in the shade in areas with hot summers. It is important to keep it well watered during the growing season. Deadhead flowers to encourage blooming. Space 9–12 inches apart.

NOTE: *Monkey flowers are popular with children, since the flowers resemble monkeys' faces.*

Description: Tubular, boldly colored red, orange, yellow, and pink spotted flowers bloom from late spring to frost on plants that have a mounded habit. The foliage is also ornamental with variegated green and lighter green foliage. It makes a nice plant for moist growing areas and does well as a container plant if kept well watered. Height: 1–1½ feet.

MIGNONETTE. SEE *RESEDA ODORATA*

MIRABILIS JALAPA

Annual
Four-o'clock, marvel of Peru

Sowing Directions: Sow seeds indoors 4–6 weeks before the last frost. Can be sown outdoors, after the last frost, where they are to grow. Established plants will self-sow.

Indoor Germination Temperature: 70–75°F

Days to Germination: 4–7

Growing On Temperature: 60°F

Garden Planting: Plant in full sun to part shade in any well-drained soil. It tolerates heat and poor soils. Space 12–18 inches apart.

Mirabilis Jalapa 'Floranova'

Description: This fast-growing plant has a bushy habit. Cheery, bright white, magenta, pink, light purple, orange, and yellow trumpet-shaped, fragrant flowers bloom throughout the summer to frost. As the common name indicates, the flowers begin to open in late afternoon and continue blooming into the evening, closing again in the morning. Plant in the border or in containers near sitting areas so that the scent of the flowers can be enjoyed. Height: 2–3 feet.

MOLUCCELLA LAEVIS

Annual
Bells of Ireland

Sowing Directions: For indoor sowing, refrigerate seeds for 1 week and then soak overnight in warm water before sowing. Sow seeds indoors 8–10 weeks before the last frost into individual peat pots that can be transplanted directly to the garden. For outdoor sowing, soak seeds overnight in warm water before sowing. For best results, sow seeds outdoors, in the late fall or early spring, directly where plants are to grow. Established plants will self-sow. Seeds need light to germinate; lightly press onto the surface; do not cover with medium.

NOTE: *These seeds like high daytime temperatures and cool nights for good germination results. Be prepared for erratic germination results with bells of Ireland.*

Indoor Germination Temperature: 80–85°F (day), 50°F (night)

Days to Germination: 12–21

Growing On Temperature: 60°F

Garden Planting: After the last frost, plant in full sun to light shade in any well-drained soil. Taller specimens may need staking. Space 1 foot apart.

Description: This very distinctive plant has erect stems that bear light green, bell-shaped, papery calyx flowers with white veination in late summer. It is prized as a dried flower (it turns a lovely straw color) and for its unusual accent color and uncommon architectural structure in beds and borders. Height: 2–2½ feet.

MONARDA DIDYMA

Perennial
Bee balm, bergamot
Zones: 3–9

Sowing Directions: Sow seeds indoors 8–10 weeks before the last frost. For best results, sow seeds outdoors, in early spring through midsummer, directly where they are to grow. Lightly cover seed with soil.

Indoor Germination Temperature: 60–70°F

Days to Germination: 10–20

Growing On Temperature: 55–60°F

Garden Planting: After the last frost, plant in full sun to light shade in a moist, well-drained soil. It prefers moist soils but will grow in dry conditions. Deadhead flowers to promote more blooms. Space 2 feet apart.

Description: A vigorous growing plant that produces scarlet red, pink, purple, or white downy looking, whorled tubular flowers in early to midsummer. The brilliantly colored flowers seem to be almost fluttering at times and are magnets for bees, butterflies, and hummingbirds. It looks best when planted in drifts in the border or massed in naturalized wildflower areas. Susceptible to powdery mildew. Height: 3–4 feet.

Recommended Cultivar: 'Panorama' produces gorgeous shades of pink, salmon, and crimson flowers that bloom from mid to late summer. Height: 3 feet.

MORNING GLORY. SEE *IPOMOEA*

MYOSOTIS SYLVATICA

Biennial
Forget-me-not, woodland forget-me-not
Zones: 5–8

Sowing Directions: Start seeds indoors 8–10 weeks before the last frost. Can be sown outdoors, in early spring, where they are to grow. Plants grown from seed sown in early spring may bloom late in the season the first year. Plants grown from seed sown in late summer will bloom the following spring. Established plants will readily self-sow. Seeds need darkness to germinate; cover seeds with medium.

> **NOTE:** *Seedlings are susceptible to damping off so take all precautions to avoid problems by using clean containers and a sterile medium. Provide good air circulation and avoid overwatering.*

TIP:
Bee balm can become invasive, so be sure to pull out unwanted new plants if you're growing it in mixed borders. Dividing and pulling out plants will also help decrease the problems of powdery mildew, since it will increase air circulation around the plants.

Indoor Germination Temperature: 68–72°F

Days to Germination: 8–14

Growing On Temperature: 55°F

Garden Planting: Plant in part shade in moist, rich, slightly acidic soil. Excellent plant for shady sites with moist to wet soils. Space 10–12 inches apart.

Description: This sweet and delicate plant with a clumping habit bears tiny blue flowers with yellow centers that bloom profusely in early to mid-spring. Use in borders, rock gardens, or naturalized wildflower or woodland areas. Height: 6–8 inches.

M. oblongata is an annual forget-me-not. Plant in full sun to part shade. Height: 12 inches.

Recommended Cultivar: 'Blue Bird' is a bushy plant with bright blue flowers. Height: 12 inches.

N

NASTURTIUM. SEE *TROPAEOLUM MAJUS*

NICOTIANA ALATA

Annual
Flowering tobacco

Sowing Directions: Sow seeds indoors 6–8 weeks before the last frost. Can be sown outdoors, after the last frost, directly where they are to grow. Mature plants will readily self-sow; however, for hybrid plant characteristics, sow new seed each year. Seeds need light to germinate; do not cover with medium.

Indoor Germination Temperature: 70°F

Days to Germination: 10–15

Growing On Temperature: 60°F

Garden Planting: Plant in light shade in moist, slightly acidic, well-drained soil. It will tolerate full sun or a site with afternoon shade in cooler climates. Deadhead flowers to promote continuous bloom. Space 1½ feet apart.

Description: The beautiful freshly scented star-shaped white, lime, pink, red, and lavender pendant flowers bloom throughout the summer and up to frost. It has a bushy habit. Use in beds and borders; compact varieties can be used for container plantings. Height: 1–4 feet.

Nicotiana alata

NOTE: *The acclaimed fragrance is best experienced with nonhybrid plants such as N. sylvestris.*

TIP: Forget-me-nots are often used as companion plants for spring-flowering bulbs. The blue flowers are stunning with yellow, white, or red tulips.

NIEREMBERGIA HIPPOMANICA

Grow as an annual
Cupflower, whitecup

Sowing Directions: Sow seeds indoors 8–10 weeks before the last frost. Can be sown outdoors, in mid-spring, directly where they are to grow. Seeds need light to germinate.

Indoor Germination Temperature: 75–80°F

Days to Germination: 6–15

Growing On Temperature: 60°F

Garden Planting: Plant in full sun to part shade, where summers are hot, in moist, well-drained soil. If the soil is poor, add fertilizer early in the growing season. Deadhead flowers to promote more blooms. Space 10 inches apart.

Description: This compact plant blooms profusely throughout the summer with dainty white, light

Nierembergia 'Purple Robe'

blue, or violet flowers with yellow centers. An excellent plant for containers, edgings, or rock gardens. Height: 6–9 inches.

Recommended Cultivar: 'Purple Robe' is wonderfully cloaked with splendid violet-blue flowers all season long until frost. Height: 6 inches.

NIGELLA DAMASCENA

Annual
Love-in-a-mist

Sowing Directions: Sow seeds indoors 6–8 weeks before the last frost into individual peat pots that can be transplanted directly into the garden. Seedlings do not like being transplanted. For best results, sow seeds outdoors, in early spring, directly where they are to grow. Successive sowings until early summer (or midsummer in areas with longer growing seasons) will give continuous blooms throughout the growing season. Lightly cover seed with soil.

> **NOTE:** *Love-in-a-mist germination rates may peak at only 50 percent; therefore, sow more seed than you think you need.*

Indoor Germination Temperature: 65–70°F

Days to Germination: 7–14

Growing On Temperature: 60°F

Garden Planting: After frost, plant in full sun in any well-drained soil. It is tolerant of dry conditions, but in poor soils, fertilize early in the season. Deadhead flowers for

neatness and more blooms. Space 1 foot apart.

Description: This erect plant with a branching habit produces blue, pink, red, and purple flowers that are cradled in a veil of lacy fernlike foliage throughout the summer. The flowers are used for fresh and dried arrangements; and the 1-inch seed heads are also popular for dried arrangements. Use the plant massed in borders. Height: 18 inches.

Recommended Cultivar: 'Persian Jewels' mix has 1½-inch-wide pink, white, red, and purple flowers. Height: 18 inches.

O

OENOTHERA MISSOURENSIS

Perennial
Evening primrose, Missouri evening primrose, Ozark sundrop
Zones: 4–9

Sowing Directions: Sow seeds indoors 8–10 weeks before the last frost in individual peat pots. Can be sown outdoors, in early spring, directly where they are to grow. Established plants will self-sow. Seeds need darkness to germinate; lightly cover seeds with medium.

Indoor Germination Temperature: 70–80°F

Days to Germination: 15–20

Growing On Temperature: 55°F

Garden Planting: After the last frost, plant in full sun to light shade in any well-drained soil. It will tolerate hot, dry conditions. Space 12 inches apart.

Description: In one of the brightest and most beautiful shades of yellow, these large cup-shaped flowers open in the evening and stay open until late morning. The flowers have a lovely soft fragrance and bloom all summer. It has a trailing habit and is best used massed in borders and rock gardens for spectacular effect. Use as an accent plant in the front of the border. Height: 12 inches.

OXYPETALUM CAERULEUM

Grow as an annual
Southern star, star of the Argentine

Sowing Directions: Sow seeds indoors 6–8 weeks before the last frost. Can be sown outdoors, after the last frost, directly where they are to grow. Cover seeds by ¼ inch with medium.

Indoor Germination Temperature: 70°F

Days to Germination: 6–10

Growing On Temperature: 60°F

Garden Planting: Plant in full sun in any well-drained soil. It will tolerate poor, dry soils. Pinch it back early in the growing season to promote fuller growth. Space 8–12 inches apart.

Description: Along with meconopsis, this plant has one of the most heavenly shades of luminous blue found in any garden plant. The sulfur blue star-shaped flowers have spots of dark blue on the waxy petals and bloom throughout the summer. The plant has dark green leaves and a spreading habit. After flowering, the plant produces attractive seedpods, which are used in dried arrangements. Plant in borders or containers. Height: 1½ feet.

Oxypetalum caeruleum
'Blue Cheer'

Recommended Cultivar: 'Blue Cheer' produces 2-inch light blue flowers that deepen to a richer blue. Tolerates hot, dry growing conditions very well. Height: 1½ feet.

P

PANSY. SEE *VIOLA* × *WITTROCKIANA*

Papaver

PAPAVER NUDICAULE

Annual
Arctic poppy, Iceland poppy
Zones: 6–8

Sowing Directions: Sow seeds indoors 6–8 weeks before the last frost into individual peat pots that can be directly transplanted into the garden. For best results, sow seeds outdoors, in early spring or late fall, directly where they are to grow. Once seeds have been sown in early spring, it is possible to extend the bloom period by doing additional sowings, spaced 5–6 weeks apart. In warmer zones, sow only in the late fall so the seedlings will develop during the cool part of the year. Established plants will self-sow. Cover seeds very lightly with medium.

Indoor Germination Temperature: 55–60°F

Days to Germination: 7–12

Growing On Temperature: 50–55°F

Garden Planting: Plant in full sun to very light shade in well-drained soil.

Papaver nudicaule 'Sparkling Bubbles' mix

It prefers a cool climate and doesn't like to be transplanted. Be sure to deadhead flowers for another bloom. Space 10–12 inches apart.

Description: The exquisite large white, pink, red, purple, orange, and salmon paperylike single or double flowers bloom in spring to early summer. In warmer southern areas, the bloom time will be confined to early spring. This plant looks especially nice clustered in a border. Height: 2 feet.

Recommended Cultivar: 'Sparkling Bubbles' mix is a vigorous grower with dazzling yellow, rose, orange, scarlet, cream, and soft pastel shades of large flowers borne on strong stems. A sensational cut flower.

PAPAVER ORIENTALE

Perennial
Oriental poppy
Zones: 2–8

Sowing Directions: Sow seeds indoors 6–8 weeks before the last frost into individual peat pots that can be transplanted directly to the garden. For best results, sow seeds outdoors, in early spring or late fall, directly where they are to grow. Since this seed is very fine, handle it carefully to ensure it is evenly sown. Plants grown from seed will bloom the second year. Established plants will self-sow. Seeds need light to germinate; do not cover with medium.

Indoor Germination Temperature: 55–60°F

Days to Germination: 7–14

Growing On Temperature: 50–55°F

Garden Planting: Plant in full sun to light shade in well-drained soil. It prefers soil that is not rich; in fact, the soil can be quite poor as long as it has good drainage. Keep it well watered during dry periods. It does not like to be transplanted. Taller plants may need some support if grown in exposed, windy areas. Space 15 inches apart.

Description: Much showier than *P. nudicaule,* this plant has fernlike foliage and large (4–6-inch) white, red, pink, and orange flowers with black centers that bloom in late spring to early summer. Height: 16–36 inches.

Papaver orientale 'Brilliant Red'

Recommended Cultivar: 'Brilliant Red' has a glorious and ravishing large scarlet flower with dark green foliage. Height: 3 feet.

PELARGONIUM × HORTORUM

Grow as an annual
Geranium, bedding geranium, standard geranium, zonal geranium

TIP: Oriental poppies die back in mid-summer. Fill in conspicuous gaps with later-growing plants or annuals.

Sowing Directions: Start seeds indoors 10–12 weeks before the last frost. It is not recommended to sow seeds directly outdoors except in Zones 9–11 where they can be sown in early spring. Make sure the seedlings get adequate light immediately after germination. Lightly cover seeds with medium.

Indoor Germination Temperature: 70–75°F

Days to Germination: 7–14

Growing On Temperature: 60–65°F

Garden Planting: Plant in full sun in rich, well-drained soil. Cultivars with variegated leaves prefer light shade. Keep it well watered throughout the growing season, especially during dry periods. If the soil isn't naturally rich, fertilize monthly throughout the growing season. Deadhead to keep it tidy and promote further blooms. Remove any yellow and spent foliage throughout the growing season. Transplant seedlings 12–18 inches apart.

> **NOTE:** *P. odoratissimum* (scented geranium), *P. tomentosum* (woolly geranium), *P. quercifolium* (oak-leaf geranium), *P. peltatum* (ivy-leaved geranium), and *P. radens* (crowfoot geranium) *are germinated in the same manner as P. × hortorum.*

Description: This popular classic plant needs no description. Flowers come in bright red, white, salmon, pink, and bicolor ball-shaped flower clusters. It blooms dependably all summer long. Use massed in flower beds, containers, and window boxes. Height: 10–20 inches.

Recommended Cultivars:

'Big Red' produces extra-large, brilliant scarlet flowers on strong stems. Height: 15 inches.

'Summer Showers' series mix is one of the best for containers, window boxes, and hanging baskets. Produces red, rose, pink, lavender, plum, and white flowers on cascading ivy-leaved foliage.

Pelargonium 'Summer Showers' series mix

PENNISETUM SETACEUM

Perennial; grow as an annual
Annual fountain grass, feathertop, crimson fountain grass
Zones: 7–11

Sowing Directions: Sow seeds indoors 8–10 weeks before the last frost into peat pots. Sow two to three seeds per pot, thinning back to two seedlings. Can be sown outdoors, in spring up to 2 months before the first fall frost, where they are to grow. Seeds need light to germinate; sow on the surface.

TIP:
Some gardeners like to soak geranium seed in warm water overnight before sowing.

Indoor Germination Temperature: 65–70°F

Days to Germination: 15–20

Growing On Temperature: 60°F

Garden Planting: Plant in full sun in any well-drained soil. Provide water during dry periods. Cut this plant back to the ground in late fall to early spring. Space 2–2½ feet apart.

Description: A lovely airy ornamental grass that produces 6-inch long spikes of crimson-tipped feathery flower plumes that are borne above arching foliage in mid to late summer (or early fall if started from seed in the spring). This plant does well in small clusters in borders or as a specimen plant for accent color and texture in mixed borders. Height: 3–4 feet in bloom.

TIP:

Perilla is most often used as an accent plant. It makes a unique contrast for other flowering plants, especially those with white or pink flowers. It makes a nice addition to the border as well.

PENSTEMON BARBATUS

Perennial
Beardtongue, penstemon
Zones: 5–9

Sowing Directions: Sow seeds indoors 8–10 weeks before the last frost. Can be sown outdoors in late summer or early spring. Seeds need light to germinate; sow on the surface and do not cover with medium.

Indoor Germination Temperature: 60–65°F

Days to Germination: 14–21

Growing On Temperature: 55°F

Garden Planting: Plant in full sun or light shade, in areas with hot summers, in fertile, acidic, extremely well-drained soil. It prefers cool cli-

mates and will not tolerate heavy, wet soils. Space 2 feet apart.

Description: This erect-growing plant has tubular red and pink flowers that bloom above dark green foliage from late spring to midsummer. Height: 1½–3 feet.

PERILLA FRUTESCENS

Annual
Perilla, false coleus, beefsteak plant

Sowing Directions: Sow seeds indoors 4–6 weeks before the last frost in individual peat pots that can be directly transplanted into the garden. Can be sown outdoors, after the last frost, where they are to grow. Mature plants readily self-sow. Seeds need light to germinate; sow on the surface and do not cover with medium.

Indoor Germination Temperature: 65–75°F

Days to Germination: 14–21

Growing On Temperature: 55–60°F

Garden Planting: After the last frost, plant in full sun to part shade in well-drained soil. Pinch back the tips of young plants for a fuller, bushier shape. Space 15 inches apart.

Description: Grown primarily for its unique dark purplish black and bronzy foliage, this plant's leaf and growth habit resemble the true coleus. Tiny white or pink flowers bloom in late summer. Height: 1½–2½ feet.

PERIWINKLE. SEE CATHARANTHUS ROSEUS (VINCA ROSEA)

P

PETUNIA × HYBRIDA

Annual
Petunia

Sowing Directions: For best results start seeds indoors 8–10 weeks before the last frost. Keep the humidity high when germinating this seed. Seeds need light to germinate; sow seeds on the surface and do not cover with medium.

Indoor Germination Temperature: 75–80°F

Days to Germination: 10–14

Growing On Temperature: 55–60°F

Garden Planting: After the last frost, plant in full sun to light shade in moist, well-drained soil. Keep it well watered during dry periods. Pinch back young seedlings for a bushier habit. Give young plants a fertilizer application in spring and early summer. Deadhead spent blooms, especially after a rain, to promote more blooms and keep it neat looking. Cut back in August to renew growth. Space 10 inches apart.

Description: An all-time summer favorite, this plant produces brightly colored white, purple, red, yellow, and bicolor trumpet-shaped fragrant flowers. Plant in flower beds, borders, window boxes, containers, and hanging baskets. Height: 1½ feet.

Recommended Cultivars:

'Purple Wave' produces a rolling wave of striking deep burgundy-purple flowers. This cultivar can spread 3–4 feet across. Do not cut back in August. Excellent for hanging baskets and other containers. Height: 4 inches.

'Total Madness' mix contains 15 different flower colors, including silver, rose, coral, blue, pink, and red. These large flowers stand up to the weather.

Petunia × hybrida 'Total Madness' mix

Phlox

PHLOX DRUMMONDII

Annual
Annual phlox, Drummond phlox, Texas pride

Sowing Directions: Start seeds indoors 6–8 weeks before the last frost in individual peat pots. Can be sown outdoors in early spring. Seeds need darkness to germinate.

Indoor Germination Temperature: 60–65°F

Days to Germination: 10–15

Growing On Temperature: 50–55°F

Garden Planting: Plant in full sun to light shade in moist, well-drained, slightly acidic soil. Keep this plant well watered during dry periods and fertilize it in early summer. Pinch back young plants for bushier habits, and deadhead blooms to encourage further flowering. Space 1 foot apart.

Description: This fast-growing native is perfect for adding bright splashes of pink, red, purple, yellow, and white star-shaped flowers. Most cultivars bloom in early summer with a second flowering in the fall. Use massed in beds and borders, as edging, or as a container plant. Height: 6–20 inches.

Phlox Drummondii 'African Sunset'

Recommended Cultivar: 'African Sunset' has a spreading habit that makes it excellent for edgings or hanging baskets. Flowers are intensely red and appear to glow. Height: 4 inches.

PHLOX SPECIES

Perennial
P. divaricata: Woodland phlox (Zones 4–9), *P. paniculata:* garden phlox (Zones 3–9), *P. subulata:* mountain pink (Zones 3–9) Perennial phlox

Sowing Directions: For best results sow seed in containers in the fall. Place the containers in a sheltered outdoor area and cover with glass. Water as necessary to keep the soil moist until the seedlings can be transplanted to their growing site in the garden in the following spring or fall. Seeds need darkness to germinate.

Phlox 'Subulata'

Germination Temperature: 60–65°F

Days to Germination: 14–30

Growing On Temperature: 55–60°F

Garden Planting: Plant in full sun to light shade in moist, rich, well-drained soil. Keep well watered during the growing season. Deadhead regularly and cut back plants almost to the ground after flowering in the fall. Space 1–2 feet apart.

Description: These species come in a wide variety of heights and habits. The star-shaped flowers come in shades of blue, pink, white, and pur-

TIP:

Plant phlox in areas where it will have good air circulation. It is vulnerable to powdery mildew (a fungal disease that causes a grayish fungal layer to form on foliage), especially in hot, humid conditions.

ple. There is a variety for almost any garden site, including beds, borders, rock gardens, containers, ground covers, and wildflower gardens.

P. divaricata: Terminal flowers (which bloom at the end tips of plant stems). Height: 15 inches.

P. paniculata: Terminal clusters of flowers bloom in mid to late summer and early fall. Height: 2–4 feet (in bloom).

P. subulata: A low-growing species with a spreading habit and bright pink, purple, or white flowers that bloom in early to midspring. Height: 3–6 inches.

PHYSALIS ALKEKENGI

Perennial
Chinese lanterns, Japanese lanterns, winter cherry, bladder cherry
Zones: 4–8

Sowing Directions: Sow seeds indoors 6–8 weeks before the last frost. Can be sown outdoors, in midspring, where they are to grow. Established plants will self-sow very easily. Seeds need light to germinate; do not cover seeds with medium.

Indoor Germination Temperature: 60–70°F

Days to Germination: 7–14

Growing On Temperature: 55°F

Garden Planting: Plant in full sun to light shade in any well-drained soil. Space 2 feet apart.

Description: Grown for its bright warm orange seedpods that look like small paper lanterns strung along the plant. Small white flowers bloom in early summer. The pods are used in fresh and dried flower arrangements. Height: 2 feet.

PHYSOSTEGIA VIRGINIANA

Perennial
False dragonhead, obedient plant, stay-in-place
Zones: 3–9

Sowing Directions: Sow seeds indoors 8–10 weeks before the last frost. Can be sown outdoors in early spring or fall. Lightly cover seeds with soil.

Indoor Germination Temperature: 65–70°F

Days to Germination: 20–25

Growing On Temperature: 55–60°F

Garden Planting: Plant in full sun to light shade in moist, well-drained soil. Keep it well watered if the soil is dry or rainfall is scant. Space 1–2 feet apart.

Physostegia virginiana

Description: The brightly colored pink, white, and purple snapdragon-

TIP:
If you are growing Chinese lanterns as an annual, pull it up completely in the fall. Plants can become invasive.

TIP:
False dragonhead can spread quickly—to the point of being pesky. To avoid invasiveness, pull up unwanted plants during the growing season.

like flowers bloom atop erect growing stems. It has a spreading habit and is best used in borders and wildflower sites. Height: 3 feet.

PLATYCODON GRANDIFLORUS

Perennial
Balloon flower, Chinese bellflower
Zones: 3–9

Sowing Directions: Sow seeds indoors 6–8 weeks before the last frost into individual peat pots that can be directly transplanted to the garden. Can be sown outdoors, in spring and up to 2 months before the first fall frost, where they are to grow. Seeds need light to germinate; do not cover with medium.

Indoor Germination Temperature: 60–70°F

Days to Germination: 7–14

Platycodon grandiflorus var. mariesii

Growing On Temperature: 55–60°F

Garden Planting: Plant in full to light shade in moist, well-drained soil. Deadhead spent flowers to prolong flowering period. Space 1–1½ feet apart.

Description: This delightful plant has light purple flowers that, just before they open to reveal their lovely star shape, resemble balloons. It blooms in mid to late summer. Plant in the front to middle of a border. Height: 2–2½ feet.

Recommended Cultivar: 'Fuji' mix is a particularly beautiful mix of white, soft pink, and purple flowers that bloom profusely from early summer to frost. Height: 12–16 inches.

POLEMONIUM CAERULEUM

Perennial
Jacob's ladder, Greek valerian, charity
Zones: 3–8

Sowing Directions: Sow seeds indoors 8–10 weeks before the last frost. Can be sown outdoors, in early spring or fall, where plants are to grow. Lightly cover seeds with medium.

Indoor Germination Temperature: 70°F

Days to Germination: 21–30

Growing On Temperature: 60°F

Garden Planting: Plant in full sun to light shade (in hot, dry areas) in any well-drained soil. Keep it well watered during dry periods. Cut it back after flowering. Space 18 inches apart.

Description: This plant's common name is a reflection of its foliage: The narrow opposite leaflets look like a ladder. It has a clumping habit and a light, airy appearance and lovely blue, pink, or white cup-shaped flowers with yellow centers that bloom in late spring to early summer. Use in borders or rock gardens. Height: 2 feet.

POLYGONATUM BIFLORUM

Perennial
Solomon's seal
Zones: 4–9

Sowing Directions: Seeds must be subjected to a cold treatment before germination. For outdoor sowing in early spring, sow seeds into containers, wrap them in plastic, and place them in a refrigerator (35–40°F) for 3 weeks. Then move the containers to a sheltered, shady outdoor spot, cover with glass, and keep the soil moist. For outdoor sowing in the fall, sow seeds in containers and sink them into the ground in a sheltered, shady area (against a north-facing wall is ideal). Cover them with glass and keep the soil moist. In early spring, lift the containers and move them to a place where they can warm to 50–55°F. Transplant seedlings grown from either method to their permanent site when they appear.

Germination Temperature: 70–75°F

Days to Germination: 30–180

Growing On Temperature: 60–65°F

Garden Planting: Plant in part to full shade in rich, moist, well-drained, acidic soil. Apply well-composted manure or a light fertilizer early in the growing season. Cut it back in the fall. Space 2 feet apart.

Description: This regal shade plant has slender arching stems that bear beautiful ovate-veined light green foliage. Small, white tubular flowers gracefully dangle all along the stems in late spring and early summer. Plant in small clusters, use as accents, or include in woodland plantings. It looks lovely hanging over hosta foliage. Height: 2–3 feet.

POLYGONUM CAPITATUM

Annual
Knotweed, polygonum, pink-head knotweed, fleeceflower

Sowing Directions: Sow seeds indoors 4–6 weeks before the last frost. Can be sown outdoors, after the last frost, where plants are to grow.

Indoor Germination Temperature: 70–75°F

Days to Germination: 21–30

Growing On Temperature: 60°F

Garden Planting: Plant in full sun or light shade in well-drained soil and keep it well watered. It can become invasive, so keep it in bounds by cutting back as needed. Space 2 feet apart.

Description: This vigorous growing, fairly dense ground cover spreads out to about 2 feet. Small spikes of white and pink flowers bloom all summer to frost. Use as an edging, in a border, or in a rock garden. Try letting it trail down the side of a container. Height: 4 inches.

POPPY. SEE *PAPAVER*

PORTULACA GRANDIFLORA

Annual
Portulaca, rock moss, moss rose, sun plant

Sowing Directions: Sow seeds indoors 6–8 weeks before the last frost into individual peat pots that can be directly transplanted into the garden. For best results, sow seeds outdoors, after the last frost, directly where they are to grow. Be sure that the seed bed is kept moist until the seedlings are established. Mature plants will self-sow. Seeds need light to germinate; lightly sow seed on surface and do not cover with medium.

NOTE: *Higher temperatures are essential for good indoor germination rates.*

Indoor Germination Temperature: 75–80°F

Days to Germination: 7–10

Growing On Temperature: 65°F

Garden Planting: Plant in full sun in dry, well-drained soil. If the soil is not naturally dry, add some sand to the planting area. It will thrive in hot climates and is very drought-tolerant. Space 1–2 feet apart.

Description: One of the brightest colored garden plants with neon fuchsia, yellow, pink, red, orange, and white single and double roselike satiny flowers that bloom all summer until the first frost. The flowers close on cloudy days and in the late afternoon. It has succulent-type foliage and will spread out horizontally to 2–2½ feet. Use in masses in beds, borders, and rock gardens. It is also good as an edging or container plant. Height: 4–6 inches.

Portulaca grandiflora 'Sundial' hybrid mix

Recommended Cultivar: 'Sundial' hybrid mix produces large, double, brilliantly colored cream, fuchsia, pink, orange, scarlet, yellow, and white flowers that stay open longer than other varieties. Height: 5 inches.

PRIMULA × POLYANTHA

Perennial
Primrose
Zones: 3–8

Sowing Directions: Sow seeds indoors 8–10 weeks before the last frost. For best results, sow freshly ripened seeds into containers in early to midsummer and place them in a protected, shady area of the garden. Put a piece of glass or clear plastic over the containers, and make sure the soil is kept moist. Transplant the seedlings their permanent site in the fall or the following spring. Seeds need light to germinate; sow on the surface and lightly sift some medium over them.

Indoor Germination Temperature: 60–65°F

Days to Germination: 21–45 (or more)

Growing On Temperature: 60–65°F

Garden Planting: Plant in part shade in a moist, acidic, well-drained soil. Keep it well watered during hot and dry spells. Space 8–12 inches apart.

Description: This delightful, old-fashioned, spring-blooming plant produces clusters of red, pink, blue, purple, orange, yellow, and white fragrant flowers that bloom atop flower stalks that rise above the foliage. All the flowers have yellow centers. Height: 10 inches.

Recommended Cultivar: 'Dwarf Jewel' mix bears 2½-inch white, yellow, pink, rose, red, blue, and purple flowers Height: 6 inches.

PULSATILLA

Perennial
Pasqueflower, spring anemone, alpine anemone
Zones: 4–8

Sowing Directions: For best results start seeds outdoors and expose them to winter temperatures. Use fresh seed as it becomes available in the summer and sow it into containers. Cover the containers with glass or clear plastic and place them outdoors in a sheltered, shady site; keep the potting mix moist. If the seeds have not germinated by the fall, keep them outside or place the containers in a cold frame over the winter. In early spring, bring the containers indoors and warm to 60–70°F until the seeds germinate. Transplant seedlings to the garden as they emerge.

Indoor Germination Temperatures: 60–70°F

Days to Germination: 30–190

Growing On Temperature: 55–60°F

Garden Planting: In spring or fall, plant in full sun or part shade (in warmer areas) in moist, fertile, slightly alkaline, well-drained soil. Keep well watered if rainfall is low. Space 1 foot apart.

Description: White, red, yellow, or purple bell- or cup-shaped flowers appear in spring to early summer. The fernlike foliage is light and finely cut. Use in rock gardens. Height: 6–12 inches.

PYRETHRUM. SEE
CHRYSANTHEMUM

TIP:
Some gardeners sow primrose seed in moist medium and then subject the covered containers to 3–4 weeks of refrigeration before warming the potting mix for germination.

R

RANUNCULUS ASIATICUS

Perennial; grow as an annual
Buttercup
Zones: 8–9

Sowing Directions: These seeds must be subjected to a cold treatment before germinating. For fall sowing, sow seeds into containers and place them outdoors in a sheltered, shady site. Cover them with glass or plastic and keep the potting mix. In early spring bring the containers indoors and warm to 50–60°F until seedlings emerge. For spring sowing, place seeds in a plastic bag with some moistened soil medium and refrigerate for 3 weeks. Then sow seeds into containers and place them outdoors in a sheltered, shady site. Cover them with glass or plastic and keep the potting mix moist until seedlings emerge.

Indoor Germination Temperature: 50–60°F

Ranunculus asiaticus 'Tecolote' mix

Days to Germination: 14–21

Growing on Temperature: 50–55°F

Garden Planting: After the last frost, plant in full sun in moist and well-drained soil. Use mulch to keep the soil moist and cool; keep the plant well watered during dry periods. Space 1 foot apart.

Description: This upright, branching plant is grown for its single or double bright yellow, gold, pink, red, rose, or white cup-shaped flowers. Use in borders or rock gardens. Blooms in late spring. Height: 1½ feet.

RESEDA ODORATA

Annual
Mignonette

Sowing Directions: Start seeds indoors 6–8 weeks before the last frost into individual peat pots that can be directly transplanted to the garden. Can be sown outdoors directly where they are to grow. Beginning in early spring and into early summer make successive sowings 3 weeks apart for continuous bloom throughout the summer. Mature plants will self-sow. Seeds need light to germinate; sow onto the surface and do not cover with medium.

Indoor Germination Temperature: 70°F

Days to Germination: 5–10

Growing On Temperature: 60–65°F

Garden Planting: After the last frost, plant in full sun to light shade in moist, well-drained soil. Keep the soil moist throughout the growing season. Space 1 foot apart.

Description: This upright plant is grown and loved for its beautifully scented white or yellow flowers that bloom in summer. Use in borders or cutting gardens. Height: 1–1½ feet.

RUDBECKIA

Perennial
R. fulgida, R. hirta
Black-eyed Susan, gloriosa daisy
Zones: 3–9

Sowing Directions: For indoor sowing, 8–11 weeks before the last frost, put seeds in a plastic bag with moistened medium and place in the refrigerator for 2–3 weeks. Sow treated seeds into containers to germinate. For outdoor sowing, in early to late spring sow seed into containers and place them outdoors in a sheltered, shady site. Keep the potting mix moist. Transplant young plants to

Rudbeckia hirta 'Indian Summer'

their garden site in the fall or the following spring. Plants grown from seed will bloom the first year if started early. Seeds need light to germinate; sow seeds onto the surface; do not cover.

Indoor Germination Temperature: 70–72°F

Days to Germination: 7–14

Growing On Temperature: 50°F

Garden Planting: After the last frost, plant in full sun in any well-drained soil. It is tolerant of a wide range of soils and can withstand dry conditions. However, for best performance, water during dry spells. If the soil is very poor, add some composted manure or fertilizer in early summer. Taller varieties may need staking in windy sites. It can become invasive, so divide as necessary to control plants. Space 2 feet apart.

Description: One of the easiest and most widely grown perennials, it has bright yellow-gold, daisylike flowers with dark brown centers that bloom freely from midsummer to early fall. Use this upright growing plant in borders and naturalized areas, including wildflower meadows. Height: 1–4 feet.

Recommended Cultivars:

R. fulgida 'Goldstrum' is a dependable cultivar with bright golden-yellow flowers. Height: 1½–2 feet.

R. hirta 'Indian Summer' has huge (6–9-inch) single and semidouble golden flowers. It is great for cut flowers and does not require staking. Height: 3–3½ feet.

TIP:
Black-eyed Susans look especially lovely when planted with ornamental grasses.

S

SALPIGLOSSIS SINUATA

Annual
Painted tongue

Sowing Directions: For best results, sow seeds indoors 8–10 weeks before the last frost into individual peat pots that can be directly transplanted into the garden. Can be sown outdoors, after the last frost, directly where they are to grow. Sow these very fine seeds onto the surface and then lightly sift a trace of potting mix or soil over them. Seeds need darkness to germinate, so cover them with newspaper or black plastic (indoors) or straw, burlap, or cheesecloth (outdoors). Remove the cover after the seedlings emerge.

Indoor Germination Temperature: 65–70°F

Days to Germination: 12–15

Growing On Temperature: 50–55°F

Garden Planting: After the last frost, plant in full sun in any well-drained soil. For best flowering, provide cool growing conditions and deadhead spent flowers. Pinch back young plants to promote more branching. Tall specimens often need to be staked. Space 12–15 inches apart.

Description: Colorful trumpet-shaped red, purple, yellow, pink, and blue flowers bloom from midsummer to frost. Use this upright plant in borders, containers, and the cutting garden. Height: 2–3 feet.

SALVIA

Grow as an annual
S. coccinea, S. farinacea, S. horminum, S. splendens

Sowing Directions: For best results, sow seeds 8 weeks before the last frost. (Start *S. farinacea* 10–12 weeks before the last frost.) Seeds need light to germinate; sow onto the surface.

Indoor Germination Temperature: 70–75°F

Salvia farninacea 'Victoria' and 'White Porcelain'

Days to Germination: 14–19 (*S. coccinea*: 18–26)

Growing On Temperature: 60°F

Garden Planting: Plant in full sun in any well-drained soil. It will tolerate very hot and humid conditions. Space 8–10 inches apart.

Description: This garden workhorse provides intense color all summer long up until frost with its upright stems spikes of bright red, purple, pink, blue, and white tubular flowers. Use it as edgings, in borders, and in containers. Height: 10–30 inches, depending on variety.

Recommended Cultivars:

S. coccinea 'Lady in Red' produces brilliant scarlet blooms along 24-inch spikes. Attractive to hummingbirds and butterflies.

S. farinacea 'Victoria' has lovely, soft, clear blue flowers and a compact habit. Height: 18 inches.

S. farinacea 'White Porcelain' has cooling white flowers. Height: 18 inches.

S. splendens 'Flare' has bright red upright flowers with dark green foliage. Excellent for edgings. Height: 20 inches.

SANTOLINA CHAMAECY-PARISSUS

Perennial
Lavender cotton
Zones: 6–9

Sowing Directions: To sow seeds indoors, first place seeds in a plastic bag with some moistened medium and chill for 2–3 weeks. Then sow treated seeds 8–10 weeks before the last frost. Can be sown outdoors, in early spring up to 2 months before the first fall frost. Lightly cover seeds with medium by about ⅛ inch.

Indoor Germination Temperature: 65–70°F

Days to Germination: 15–20

Growing On Temperature: 60°F

Garden Planting: In early to mid spring, plant in full sun in any very well-drained soil. It will grow in poor or dry soil, as long there is good drainage. Shear back plants after flowering to maintain a nice shape and promote more blooms. Space 20 inches apart.

Description: This drought-tolerant plant has finely cut, silvery gray, fernlike foliage and small, round, bright golden yellow flowers that are excellent for fresh or dried arrangements. It has a rounded habit, and the foliage is nicely aromatic. It is often used in herb gardens as edging and is nice for mixed borders and rock gardens. Height: 1½–2 feet.

SANVITALIA PROCUMBENS

Annual
Creeping zinnia, trailing sanvitalia

Sowing Directions: Sow seeds indoors 4–6 weeks before the last frost into individual peat pots that can be directly transplanted to the garden. Can be sown outdoors after the last frost. Seeds need light to germinate; do not cover with medium.

Indoor Germination Temperature: 70°F

Days to Germination: 7–10

Growing On Temperature: 60°F

Garden Planting: After the last frost plant in full sun in well-drained soil. It will tolerate dry soils; however, provide extra water during dry periods. Space 6–9 inches apart.

Description: This charming annual has bright, golden yellow mini-sunflowerlike flowers that bloom from early summer to frost. Its trailing habit that makes it wonderful for hanging baskets and other containers. Height: 6 inches.

SAPONARIA OCYMOIDES

Perennial
Rock soapwort, soapwort
Zones: 4–8

Sowing Directions: Sow seeds indoors 8–10 weeks before the last frost to individual peat pots that can be directly transplanted into the garden. Can be sown outdoors in early spring or fall. Seeds need darkness to germinate; cover seeds completely with soil.

Indoor Germination Temperature: 70°F

Days to Germination: 8–12

Growing On Temperature: 55–60°F

Garden Planting: After the last frost, plant in full sun in sandy, well-drained soil. Shear back plants after flowering to maintain a neat appearance and prevent invasiveness. Space 12–15 inches apart.

Description: Masses of small, pink and rose flowers that bloom in late spring to early summer contrast nicely with this plant's dark green succulent foliage. Plant in the rock garden, over walls, or in borders as an edging; it has a horizontal spread of about 3 feet. Height: 9–10 inches.

SCABIOSA ATROPURPUREA

Annual
Pincushion flower

Sowing Directions: For best results, start seeds indoors 4–5 weeks before the last frost. Can be sown outdoors, after the last frost, where

they are to grow. Very lightly cover seeds with medium.

Indoor Germination Temperature: 65–70°F

Days to Germination: 10–12

Growing On Temperature: 50–55°F

Garden Planting: In the spring, plant in full sun in moist, rich, well-drained alkaline soil. Deadhead spent flowers. Tall specimens may require staking. Space 1 foot apart.

NOTE: *The perennial* S. caucasica *can be sown directly into the garden in early spring or early fall.*

Description: This plant's common name comes from the rounded puffy appearance of its purple, blue, lilac, and white 3-inch flowers that bloom in midsummer. It makes an excellent cut flower and looks best planted in drifts or masses in beds and borders. Height: 2–3 feet.

Recommended Cultivar: 'Giant Imperial' mix produces outstanding 3-inch, ball-shaped blue, white, rose, pink, salmon, crimson, and lavender flowers that bloom all summer. Height: 3 feet.

SCHIZANTHUS

Annual
Poor man's orchid

Sowing Directions: Sow seeds indoors 10–12 weeks before the last frost. In Zones 9–11, can be sown outdoors in late fall or early spring. Seeds need darkness to germinate;

TIP:
Poor man's orchid can be grown in a cool greenhouse in the winter for cut flowers on snowy days.

Schizanthus × wisetonensis 'Angel Wings' mix

cover seeds with ⅛ inch of medium.

Indoor Germination Temperature: 60–70°F

Days to Germination: 20–25

Growing On Temperature: 50°F

Garden Planting: After the last frost, plant in full sun in rich, moist, acidic soil. It will thrive only in areas with cool summers. In hot climates, water frequently; but it will likely burn out during the summer. Pinch back its tips when young for a bushier shape. Space 1 foot apart.

Description: This unusual annual has an upright, bushy habit, fernlike foliage, and masses of lovely, small red, pink, and white orchidlike flowers that bloom in spring and summer. Use in borders, as an edging, or in containers. Height: 1–2 feet.

SEDUM

Perennial
Stonecrop
Zones: 4–9

Sowing Directions: Sow seeds indoors 6–8 weeks before the last frost. Can be sown outdoors in containers in the fall. The containers should be placed in a protected outdoor area to overwinter. In the early spring, bring the container indoors to germinate at 70–80°F. Most gardeners purchase plants or propagate by stem cuttings.

Indoor Germination Temperature: 70–80°F

Days to Germination: 15–30

Growing On Temperature: 60–65°F

NOTE: *There are many sedum species. Some are quite low growing and can be used as ground covers.*

Sedum 'Autumn Joy'

Garden Planting: After the last frost, plant in full sun in sandy well-drained soil. It tolerates heat and drought conditions. Space 1 foot apart.

Description: This plant has a rounded growth habit and fleshy, succulent, light green to silvery green leaves. Flowers bloom in flat-topped clusters of tiny white, red, and pink flowers in late summer until the first frost. Plant in borders, rock gardens, or masses and use in hot, dry sites. Height: 2 feet.

Recommended Cultivar: 'Autumn Joy' develops light green flowers that turn to a rich pink and then a dark burgundy in late summer and fall. Its habit is very tidy, and the foliage is a lovely light grayish green. One of the best perennials for American gardens. Height: 2 feet.

SEMPERVIVUM TECTORUM

Perennial
Hen-and-chickens, houseleek
Zones: 5–9

Sowing Directions: Sow seeds indoors anytime in early spring up to early summer.

Indoor Germination Temperature: 75–85°F

Days to Germination: 15–35

Growing On Temperature: 65–70°F

Garden Planting: Plant in full sun and poor, gravelly or gritty, dry, very well-drained soil. Cut back flower stalks after flowering. Space 8–12 inches apart.

Description: This fleshy, succulent-leaved, low-growing plant is shaped like an upturned blooming flower and is primarily grown for its rosette shape and foliage. The foliage is an attractive light green to bronzy red. Its 12–15-inch flower stems bear white, pink, purple, and red flowers that bloom in midsummer. Use this as edging, in between pavement or stepping stones, in rock gardens, or in trough gardens. Height: 3–4 inches.

SIDALCEA MALVIFLORA

Perennial
Checkerbloom, false mallow, checkermallow, prairie mallow
Zones: 5–9

Sowing Directions: Sow seeds indoors 6–8 weeks before the last frost. Can be sown directly outdoors in early spring. Cover seeds with 1/8 inch of medium.

Indoor Germination Temperature: 55°F

Days to Germination: 14–30

Growing On Temperature: 50°F

Garden Planting: After frost plant in full sun or light shade in moist soil. Cut back plants after flowering to encourage a second bloom period. Provide extra water during dry periods. Tall specimens may need staking. Space 1 foot apart.

Description: This upright growing plant produces pink, purple, and white hollyhock-type flowers that bloom in the summer and has attractive, deeply cut foliage. Use in the middle or back of the border. Height: 2–4 feet.

SNAPDRAGON. SEE *ANTIRRHINUM MAJUS*

SOLIDAGO

Perennial
Goldenrod
Zones: 4–8

Sowing Directions: Sow seeds indoors in late winter. Can be sown outdoors, in early spring, directly where plants are to grow. Mature plants will self-sow if flowers are left intact. Usually propagated by division. Lightly cover the seeds with medium.

Solidago 'Cloth of Gold'

Indoor Germination Temperature: 60–65°F

Days to Germination: 14–28

Growing On Temperature: 55–60°F

Garden Planting: In the spring, plant in full sun to light shade in well-drained soil. Space 1½–2 feet apart.

Description: This vigorous grower has a clumping habit and is one of the best perennials to add bright golden yellow color to the gardens in late summer and fall. Tiny yellow clusters of flowers bloom atop arching branches. Use in the back of borders and in wildflower plantings. Height: 3–5 feet.

STACHYS BYZANTINA

Perennial
Lamb's ears, woolly betony
Zones: 4–9

Sowing Directions: Sow seeds indoors 6–8 weeks before the last frost. Can be sown outdoors where they are to grow in midspring and

up to 2 months before the first fall frost. Mature plants self-sow.

Indoor Germination Temperature: 70°F

Days to Germination: 8–16

Growing On Temperature: 60°F

Garden Planting: After the last frost, plant in full sun in well-drained soil, which is important because this plant is prone to rotting problems. It prefers a cool growing climate and low humidity. To keep the plants looking attractive, deadhead flowers and flower stalks after blooming and remove all yellow and browning leaves. Space 1–1½ feet apart.

Description: This plant is grown for its silvery grayish blue foliage, which has a soft, woolly texture. Purple to pinkish flowers bloom atop 6-inch flower stalks in early summer. Use for edgings, ground cover (it will spread about 2 feet), and as a contrast plant for brightly colored flowers. Height: 6–12 inches.

STOCK. SEE MATTHIOLA INCANA

STOKESIA LAEVIS

Perennial
Stokes' aster
Zones: 5–11

Sowing Directions: For best results, sow seeds indoors 8–10 weeks before the last frost. Plants started from seed will bloom the first year if sown this early. Can be sown outdoors in early spring and up to 2 months before the first fall frost. Lightly cover seeds with medium.

TIP:
For border plantings, use cultivated varieties of goldenrod that are less apt to become invasive, like "Cloth of Gold" or 'Golden Shower'.

TIP:
Some gardeners find the flowers unattractive and remove the flower stalks before they bloom.

Indoor Germination Temperature: 70°F

Days to Germination: 21–30

Growing On Temperature: 60°F

Garden Planting: In the spring or fall, plant in full sun to light shade in rich, well-drained soil. It will not grow in heavy, wet soils.

Description: Beautiful blue, purple, pink, and white asterlike fringed flowers bloom from midsummer to early fall. It looks best planted in the front of a border. Height: 1–2 feet.

SWEET ALYSSUM. SEE *LOBULARIA MARITIMA*

SWEET PEA. SEE *LATHYRUS ODORATUS*

TIP:
Marigolds are often planted in vegetable gardens to help repel damaging insects.

T

TAGETES

Annual
Marigold, French marigold, African marigold, Mexican marigold, signet marigold

Sowing Directions: Sow seed indoors 6–8 weeks before the last frost. Can be sown outdoors, after the last frost, where they are to grow. Lightly cover seeds with medium.

Indoor Germination Temperature: 70–75°F

Days to Germination: 6–8

Growing On Temperature: 65–70°F

Tagetes 'Jaguar'

Garden Planting: Plant in full sun in any well-drained, slightly acidic soil. Provide fertilizer early in the growing season and reapply in midseason. Keep it well watered throughout the season. Pinch back the taller-growing types for a neater and stronger-branching plant. Deadhead spent flowers to promote more blooms. Space 6 inches (for dwarfs), 1 foot (medium), or 1½–2 feet (tall) apart, depending on variety.

NOTE: *Going by the number of gardeners who grow it, the marigold, and not the rose, could be considered the national flower! Because there are many lovely new varieties, interest in the classic marigold hasn't faded.*

Description: This very easy-to-grow plant provides bright yellow to gold, orange, and white flowers all summer until the first frost. Many types of flowers are available, ranging from large and rounded to daintily small. Used massed in beds or in borders or plant in containers and rock gardens. Height: 6–36 inches.

Recommended Cultivars:

'French Vanilla' hybrid produces creamy white, 3-inch, rounded flowers. Height: 2 feet.

'Jaguar' produces yellow flowers with maroon streaks around the flower centers. Height: 10 inches.

'Lady' hybrid series: 'First Lady' is one of the best cultivars ever. It produces masses of gorgeous yellow flowers all season long. Height: 16–20 inches.

'Nugget Supreme Yellow' is a very weather-tolerant cultivar with 2-inch, bright yellow ruffled flowers. Height: 1 foot.

THALICTRUZM AQUILEGIFOLIUM

Perennial
Meadow rue, columbine meadow rue
Zones: 5–9

Sowing Directions: For best results, sow seed into containers in midsummer to fall and place them outdoors in a shady, protected site. Keep the potting medium moist. Transplant the seedlings in the fall or the following spring. Seeds can be sown, in late summer to early fall, where they are to grow. Cover seeds with about 1/8 inch of medium.

Germination Temperature: 70–75°F

Days to Germination: 15–28

Growing On Temperature: 60–65°F

Garden Planting: Plant in full sun to part shade in moist, acidic, well-drained soil. In hot-weather areas, plant in light to part shade. Keep this plant well watered during dry periods. Space 2 feet apart.

Description: This truly graceful and elegant perennial has an open habit and columbinelike airy foliage and powder-puff rose, pink, and white-colored flowers that bloom from early to midsummer. Use in the border, in lightly shaded naturalized areas, or in woodland areas. Height: 2–3 feet.

THUNBERGIA ALATA

Annual
Black-eyed Susan vine

Sowing Directions: For best results, sow seeds indoors 6–8 weeks before the last frost into individual peat pots that can be directly transplanted to the garden. In warmer zones with long growing seasons, seeds can be sown outdoors, in early spring, where they are to grow. Lightly cover seeds with medium.

Indoor Germination Temperature: 70–75°F

Days to Germination: 6–12

Growing On Temperature: 60°F

Garden Planting: Plant in full sun to very light shade in moist, rich, well-drained soil. If the soil is poor, apply

Thunbergia alta 'Susie' mix

a fertilizer and mulch. Plant near some kind of support, trellis, or fence and train the vine to climb as it grows. Keep it well watered throughout the season. Pinch or cut back to control the vine's spread. Space 1–1½ feet apart.

Description: This climbing plant has white, orange, and yellow flowers with dark brown centers. It looks lovely trailing over the sides of a container or hanging basket. Height: 6 feet in one season.

Recommended Cultivar: 'Susie' mix contains a colorful selection of white, orange, and yellow flowers. Height: 5 feet in one season.

TITHONIA ROTUNDIFOLIA

Annual
Mexican sunflower

Sowing Directions: For best results, sow seeds indoors 6–8 weeks before the last frost. Can be sown outdoors, after the last frost, directly where they are to grow. In mild climates, sow directly outdoors in early spring. Sow seeds onto surface and lightly press them into the medium.

Indoor Germination Temperature: 70°F

Days to Germination: 5–14

Growing On Temperature: 60°F

Garden Planting: Plant in full sun in any well-drained soil. It will tolerate hot, dry conditions. Tall specimens may need some support as they grow. Space 2 feet apart.

Description: Stunning reddish orange daisy- or single dahlialike

flowers bloom atop plants that have a shrubby habit with attractive dark green foliage. Use this in the back of borders or as an annual hedge or screen. Height: 3–6 feet.

Recommended Cultivar: 'Torch' is an award-winning cultivar that produces 3½-inch glowing orange flowers. Height: 4–6 feet.

TORENIA FOURNIERI

Annual
Wishbone flower

Sowing Directions: For best results, sow seeds indoors 10–12 weeks before the last frost. Seeds need light to germinate; sow onto the surface and do not cover.

Indoor Germination Temperature: 70°F

Days to Germination: 7–15

Growing On Temperature: 55–60°F

Garden Planting: After frost, plant in part to full shade in rich, moist, well-drained soil. Soil must be kept moist throughout the growing season. It can be grown in full sun only in areas with cool summers. Space 8–10 inches apart.

Torenia Fournieri 'Happy Faces' hybrid mix

Description: This low-growing, shade-tolerant plant has combinations of white, purple, blue, pink, and yellow tubular flowers. Use it as a bedding or container plant or to add color in a shady border. Height: 8–12 inches.

Recommended Cultivar: 'Happy Faces' hybrid mix produces plum, blue, pink, and white bicolored flowers. Each flower is uniquely colored! Height: 8 inches.

TRACHYMENE COERULEA

Annual
Blue lace flower, laceflower, didiscus

Sowing Directions: Sow seeds indoors 8–10 weeks before the last frost into individual peat pots that can be directly transplanted to the garden. Can be sown outdoors, after the last frost, where they are to grow. Seeds need darkness to germinate. Cover indoor germination pots with newspaper or burlap and lightly cover outdoor seeds with soil.

Indoor Germination Temperature: 70°F

Days to Germination: 15–25

Growing On Temperature: 60°F

Garden Planting: Plant in full sun in any well-drained soil. Be sure to keep it watered during dry periods. Pinch back young plants for bushier growth. Tall specimens may need some staking. Space 10–12 inches apart.

Description: This upright growing plant has attractive, finely divided foliage and masses of tiny blue to light pink and white flowers that create an airy, lacy Queen Anne's lace–type flower head. If grown in cool climates, it will bloom in early summer. Use in borders and containers. Height: 2–2½ feet.

Recommended Cultivar: 'Lacy' series produces 2½-inch pink, white, and blue flowers. Height: 2 feet.

TRADESCANTIA VIRGINIANA

Perennial
Spiderwort
Zones: 4–9

Sowing Directions: Sow seed indoors 6–8 weeks before mid-spring. For best results, sow seed directly outdoors in early spring through midsummer. Established plants will self-sow. Lightly cover seeds with medium.

Indoor Germination Temperature: 70°F

Days to Germination: 10–30

Growing On Temperature: 60°F

Garden Planting: Plant in full sun to part shade in rich, moist, well-drained, acidic soil. Keep it well watered and cut it back after flowering to promote a second blooming period. Space 18–24 inches apart.

Description: This clumping plant has long, strappy leaves and royal purple three-petaled flowers that bloom almost all summer long. Use as edgings, in front of the border, and in naturalized shady sites. Height: 2 feet.

TIP:
Spiderwort can be invasive, so divide plants as necessary to keep them contained.

TROPAEOLUM MAJUS

Annual
Nasturtium

Sowing Directions: Before sowing, scarify seeds by nicking with a sharp knife or rubbing with a nail file or piece of sandpaper. Sow scarified seeds indoors 2–3 weeks before the last frost into individual peat pots that can be directly transplanted to the garden. For best results, sow seed outside, after frost, directly where it is to grow in the garden. Seeds need darkness to germinate; cover with medium by ½ inch.

Indoor Germination Temperature: 65–70°F

Days to Germination: 10–14

Growing On Temperature: 65°F

Garden Planting: After frost, plant in full sun to light shade in a poor, dry, well-drained, sandy soil. It thrives in

Tropaeolum majus 'Alaska' mix

dry and rather infertile soil, and if soil is too rich, flowering is poor. Deadhead flowers to increase blooms and keep plants neat looking. Space 1 foot apart.

Description: This popular and very easy-to-grow annual has distinctive circular foliage and brightly colored red, yellow, and orange single or double fragrant flowers. It is available in either a bushy or a vining habit. The vine-type is especially delightful spilling out of containers or growing up a trellis. Use bush type in beds, borders, edgings, and containers. Height: 20 inches (bush), 3–6 feet in one season (vine).

Recommended Cultivar: 'Alaska' mix is a vigorous grower with marbleized green and white foliage and yellow, crimson, orange, salmon, and cherry red flowers. A bushy mounded habit with a spread of 20–40 inches. Height: 8 inches.

V

VENIDIUM FASTUOSUM

Annual
Cape daisy, monarch of the veldt

Sowing Directions: For best results sow seeds indoors 8–10 weeks before the last frost. In more southern zones, seeds can be sown outdoors, in early spring after frost, directly where they are to grow. Seeds need light to germinate. Sow seeds onto surface and press lightly into soil; do not cover.

Indoor Germination Temperature: 65–70°F

Days to Germination: 15–25

Growing On Temperature: 60°F

Garden Planting: Plant in full sun in a light well-drained soil. It will tolerate and even thrive in hot, dry sites; do not overwater. A good plant for tough locations. Space 1 foot apart.

Description: The flowers, which resemble mini sunflowers, are sunny orange, yellow, white, or red with black centers that bloom all summer to frost. Use in borders, cutting gardens, and containers. Height: 1½–2 feet.

VERBASCUM

Biennial; perennial
Mullein
Zones: 6–9

Sowing Directions: Sow seeds indoors 6–8 weeks before mid-spring. For best results, sow seeds outdoors, in early spring up to 2 months before the first fall frost, directly where they are to grow. Mature plants self-sow. Lightly cover seeds with medium.

Indoor Germination Temperature: 55–60°F

Days to Germination: 12–28

Growing On Temperature: 50–55°F

Garden Planting: In early to mid spring, plant in full sun in any well-drained, alkaline soil. Space 2 feet apart.

Description: This unique plant adds considerable architectural interest and quality to the border with its pronounced vertical habit. Its foliage is grayish silvery green and it produces bright yellow flowers that bloom all along the 3–6-foot flower stems from midsummer to early fall. Use in the back of borders.

Recommended Cultivar: 'Southern Charm' hybrid produces beautiful, soft 1-inch lavender, cream, and dusty rose flowers. This plant will bloom the first year if seed is started in February or March. Perennial in Zones 5–8.

VERBENA × HYBRIDA

Annual
Annual verbena, garden verbena

Sowing Directions: Before sowing, chill the seeds for 1 week in the refrigerator. For best results, sow treated seed indoors 10–12 weeks before the last frost. Seeds need darkness to germinate. Lightly cover seeds with potting mix and then cover the germination containers with newspaper or black plastic.

Indoor Germination Temperature: 65–70°F

Days to Germination: 10–20

Growing On Temperature: 55–60°F

Garden Planting: Plant in full sun in fertile, well-drained, acidic soil. It can withstand dry conditions but prefers extra water during dry periods. Deadhead to promote more blooms. Space 1½–2 feet apart.

Description: This low-growing plant produces clusters of glowing purple, rose, pink, white, red, salmon, yellow, and blue richly colored flowers. It can have either an upright or a spreading habit. The spreading plant will grow up to 2 feet horizontally and is attractive when allowed to spill over contain-

TIP:
Keep the germination mixture a little on the dry side and take precautions to avoid damping off problems.

er plantings. Plant the upright variety in borders, rock gardens, or in containers. Height: 8–12 inches.

NOTE: *Germination rates for verbena seeds can be low, sometimes as little as 35–50 percent.*

Recommended Cultivar: 'Peaches & Cream' produces a refreshing light color mixture of soft pink blush and apricot shades that bloom from early summer to frost. Height: 8 inches.

Verbena 'Peaches & Cream'

VERONICA SPECIES

Perennial
V. spicata: spike speedwell (Zones 4–9), *V. Teucrium*: Hungarian speedwell (Zones 4–11)

Sowing Directions: Sow seeds indoors 6–8 weeks before the last frost. Can be sown outdoors, in early spring to early summer, where plants are to grow. Seeds can also be sown outdoors in containers that are placed in a sheltered, shady site in early spring. Seedlings can be transplanted into the garden site in the fall or following spring. Seeds need light to germinate; sow onto the soil surface.

Indoor Germination Temperature: 65–75°F

Days to Germination: 16–25

Growing On Temperature: 55°F

Garden Planting: Plant in full sun to light shade in any well-drained soil. Keep well watered during dry periods. Deadhead flowers during the season and cut back plants in late fall. Space 15–18 inches apart.

Description: This plant, with its clumping, branching habit, is noted for its flowers in rich shades of blues and lavenders, but white, pink, and rose flowers are also available. The flowers bloom on erect stems.

V. spicata blooms from late spring to midsummer. Height: 8–36 inches.

V. Teucrium blooms from late spring to early summer. Height: 12–18 inches.

Recommended Cultivar: 'Sightseeing' mix produces white, pink, and blue flowers that bloom from early to late summer. Height: 28 inches. Perennial in Zones 4–9.

Viola

VIOLA TRICOLOR

Perennial
Johnny-jump-up, wild pansy
Zones: 5–9

Sowing Directions: For best results, sow seeds indoors 10–12 weeks before the last frost. Plants grown from seed started this early bloom the first year. Can be sown outdoors in late fall or early spring. Mature plants

TIP:
Johnny-jump-ups are naturally short lived, so many gardeners treat them as an annual.

will self-sow readily. Seeds need darkness to germinate; cover seeds with medium to a depth of ¼ inch.

Indoor Germination Temperature: 65–70°F

Days to Germination: 7–14

Growing On Temperature: 50–55°F

Garden Planting: Plant in part shade or full sun (only in areas with cool summers) in rich, moist, well-drained, acidic soil. Mulch to keep the soil cool and moist. Space 9 inches apart.

Viola tricolor

Description: This small pansylike plant produces purple, yellow, and white flowers that bloom in early to late spring. Use it in containers, borders, and edgings. Height: 8 inches.

VIOLA × WITTROCKIANA

Grow as an annual
Pansy, horned violet

Sowing Directions: Before sowing, place seeds in a plastic bag with some moistened medium and refrigerate for 1 week. Sow treated seeds indoors 10–12 weeks before early to midspring. Seeds need darkness to germinate. Very lightly cover seeds with medium, and then cover the container with newspaper or black plastic.

> **NOTE:** *Pansies can be grown in full sun only in cool climates. The full heat of summer in all but the coolest growing climates will reduce or eliminate flowering.*

Indoor Germination Temperature: 65–75°F

Days to Germination: 7–10

Growing On Temperature: 50–55°F

Garden Planting: In early to midspring when a light frost is still possible, plant in part shade and mulch to keep the soil moist and cool. Provide a light fertilizer feeding after planting. Deadhead spent flowers to promote more blooms and keep plants tidy. Space 8–12 inches apart.

Viola × Wittrockiana 'Blues Jam'

Description: This well-loved cheerful spring harbinger comes in a wide range of jewellike red, burgundy, blue, yellow, orange, black, and white bicolor flowers that will bloom in spring and early summer in northern gardens and in the fall and early spring in southern gardens. Use

TIP:

Be sure to sow seeds into individual pots. Sometimes transplanting can cause double-flowering seedlings to revert to single-flowering types.

in beds and container plantings and as a cool-weather edging or rock garden plant. Height: 4–12 inches.

Recommended Cultivars: These two cultivars have been especially bred for heat tolerance. They are ideal for container plantings. Height: 6–7 inches.

'Blues Jam' produces a mixture of beautiful shades of blue and white bicolor flowers.

'Rosy Cheeks' produces an assortment of delightful and velvety shades of pink, rose, and white bicolor flowers.

X

XERANTHEMUM ANNUUM

Annual
Immortelle

Sowing Directions: For best results, sow seeds indoors 3–4 weeks before the last frost into individual peat pots that can be directly transplanted to the garden. In warm zones with long growing seasons, the seed can be sown outdoors directly where they are to grow. Lightly cover seeds with medium.

Indoor Germination Temperature: 70°F

Days to Germination: 10–15

Growing On Temperature: 60°F

Garden Planting: After frost plant in full sun in dry, well-drained soil. It will tolerate heat and drought conditions quite well. Taller specimens may need some staking or other support to keep them upright. Space 12–15 inches apart.

Description: The straw-textured white, pink, rose, red, and purple 1½-inch single or double flowers bloom from midsummer to frost. Use it in beds, borders, or cutting gardens. It is an excellent dried flower. Height: 2 feet.

Z

ZINNIA

Annual
Z. angustifolia, Z. elegans
Zinnia, youth and old age

Sowing Directions: For best results, sow seeds indoors into peat pots that can be transplanted to the garden. Start *Z. elegans* 4–5 weeks and *Z. angustifolia* 8–10 weeks before the last frost. In areas with long growing seasons, seed can be sown outdoors, after frost, where plants are to grow.

Indoor Germination Temperature: 70°F

Days to Germination: 4–9

Growing On Temperature: 60°F

Garden Planting: Plant in full sun in rich well-drained soil. Provide a light fertilizer feeding early and midway through the growing season. Pinch back young plants early in the season for a fuller habit and deadhead flowers to promote more blooms and keep plants neat. Space 8–12 inches apart.

NOTE: *Zinnias can be susceptible to mildew problems. Look for seeds and plants bred for mildew-and disease-resistance and avoid wetting foliage when watering plants. These plants usually don't need much extra water, except during prolonged dry spells. Zinnias will grow fine in hot, dry conditions.*

Description: This is a familiar, old-fashioned face in the garden. Its erect stems bear splashes of brightly colored single or double flowers that bloom all summer through frost. The shape of the flowers vary and include chrysanthemum, dahlia, and ball shapes. Use them in borders, massed in beds, as edgings, or in containers.

Z. angustifolia is a low-growing plant with white or orange daisylike flowers. Height: 12–14 inches.

Z. elegans comes in almost all colors imaginable, except blue. Height: 8–36 inches.

Zinnia 'Pinwheel' mix

Recommended Cultivars:

'Pinwheel' series has been bred to be mildew-resistant and is recommended for hot, humid climates. The 3½-inch daisylike flowers come in cherry, orange, salmon, rose, and white. Colors are available individually or in a mix. Height: 12 inches.

'Splendor' hybrid series is available in individual and mixed scarlet, orange, yellow, and pink 5-inch semiruffled flowers. An excellent series for garden color and cut flowers. Height: 22 inches.

'Star White' is a must-have garden plant with white daisylike flowers with yellow-orange centers that bloom abundantly. It's a wonderful edging and container plant. Height: 14 inches.

EARLIEST
WHITE RADISH

ENCYCLOPEDIA OF
VEGETABLES *

*Listed by common name

NOTE: *The date to maturity is an estimate of when that plant's edible features are ready to be used or harvested. Unless otherwise noted, it refers to the number of days after a plant has been transplanted into the garden —and not the number of days after the seed was first sown—to harvest. Depending on the various cultivars and hybrids, the number of days to maturity for each plant type can vary significantly.*

ARTICHOKE, GLOBE

Perennial
Cynara scolymus
Zone: 8

Sowing Directions: Before sowing, soak the seeds overnight in warm water. Sow seeds indoors 6–8 weeks before the last frost. Space seeds 2 feet apart within the rows, and space the rows 4 feet apart. Lightly cover seeds with soil. Thin so seedlings are spaced 5 feet apart within the rows.

Indoor Germination Temperature: 70–75°F

Days to Germination: 18–21

Growing On Temperature: 60–65°F

Garden Needs: Plant likes a frost-free climate with cool and moist growing conditions. Sow seeds in full sun in loose, fertile, well-drained soil. Fertilize the seedlings at planting out time and again every 4 weeks. Be sure to keep the plants very well watered.

Days to Maturity: If seed is started early enough, plants may produce a crop the first year; however they usually do not produce a crop until the second year.

Harvesting Tips: The flower buds are the edible parts of this plant. The buds are ready to harvest when they are young, fairly tight, and about 4–5 inches in length. To harvest, use a sharp knife to cut off the bud off, leaving about 1 inch of stem attached to its base.

Recommended Cultivar: 'Imperial Star' hybrid will produce an abundant crop of mild, sweet-tasting flower buds the first year from seed.

ARUGULA, ROQUETTE

Eruca vesicaria subsp. *sativa*

Sowing directions: Sow seeds directly outdoors when soil can be worked in early spring. Sow evenly along the row to a depth ¼ inch. Space rows 2 feet apart. Thin seedlings to 8–10 inches apart. Succesive sowings can be made two weeks apart until late spring. Another sowing can be made in late summer or early fall in warmer climates for a fall crop.

Days to Germination: 5–7

Garden Needs: Arugula is very easy to grow spring or fall crop. Plant in a full-sun location with rich, loose, well-drained soil. Keep evenly moist and add a little organic fertilizer to planting area if soil is poor. Arugula grows well in containers.

Days to Maturity: 35–45

Harvesting Tips: Leaves have a spicy

peppery taste and are excellent for adding zest to salads. Plants are very fast growing. Leaves can be picked individually or in whole bunches when they reach 2–3 inches in length. Younger leaves taste best, so keep picking young leaves as they grow. Stop picking leaves once plants flower since leaves will become bitter-tasting.

ASPARAGUS

Perennial
Asparagus officinalis
Zone: 3

Sowing Directions: Asparagus is most often planted in the garden as root cuttings. However, seeds can be sown with successful results. Soak seeds overnight before sowing. Start seeds indoors in late winter or 12–14 weeks before the last frost date. Sow seed to a depth of ½ inch and space or thin seeds to 1 inch apart in rows at least 2 inches apart within the germination container. Use germination containers at least 4 inches deep so seedlings do not need to be transplanted before going into the garden site. Seeds can also be directly sown outdoors after the last frost when soil has warmed up a bit.

Indoor Germination Temperature: 70–75°F.

Days to Germination: 14–21

Garden Needs: Select a sunny site where plants can grow undisturbed for many years. Before sowing or planting out seedlings, the soil must be very well prepared. Plants need a rich, loose, well-drained, slightly acidic soil with lots of organic matter incorporated at least 12 inches or more into the soil. Thin or transplant seedlings 12 inches apart within rows and space rows 3 feet apart. Keep plants well watered and weeded. Lightly fertilize in the spring and again in midsummer. Allow foliage to turn brown and naturally die back in the fall before cutting it back. Rows can be mulched over the winter for extra protection.

Harvesting Tips: Asparagus is harvested beginning in the spring and continues through early summer. Be patient, since those delicious spears will not be ready to be harvested for 3 years from time of sowing. Plants are harvested by using a sharp knife to cut spears off at the base just above soil level when spears reach at least 6–8 inches in height.

Recommended Cultivar: 'Mary Washington' is a classic seed variety resistant to asparagus rust problems. If started indoors the first year, plants can be lightly harvested beginning in the second year, with normal harvests beginning in the third year.

GREEN BEAN, SNAP BEAN, STRING BEAN, WAX BEAN, KIDNEY BEAN, FRENCH BEAN, PINTO BEAN, AND LIMA BEAN

Phaseolus vulgaris, P. limensis

Sowing Directions: Sow seeds directly outdoors after the last frost and the soil has warmed up. Space seeds 2 inches apart in rows, and space rows 2 feet apart for bush types and 3 feet apart for climbing varieties.

Successive sowing of bush beans can be done every 2–3 weeks after the initial sowing up until midsummer in northern regions and until late summer in southern zones. Cover the seeds with 1–2 inches of soil. Thin the seedling to 6 inches apart within the rows. (Thin bush limas to 4 inches and pole limas to 8 inches apart.)

Days to Germination: 6–15

'Kentucky Wonder' pole bean

Garden Needs: Sow the seeds in full sun in loose, rich, well-drained soil. When the seedlings are a few inches tall, fertilize them with a light, complete fertilizer and give them another light feeding in 3–4 weeks. Keep them well watered, especially during dry periods.

Days to Maturity: Bush, 45–75; pole, 50–90

Harvesting Tips: Keep beans well picked since regular harvesting will encourage more bean pods to develop. Pick them when the pods get a little swollen, yet are still tender. For shell beans and dry beans, allow the pods to fully dry (on a screen is preferable) and crack before removing the seeds from the pods. French beans should be picked when they are the thickness of a wooden matchstick and no larger than a pencil. They mature quickly; pick them every day, or they'll get tough. Harvest limas when the pods swell and show bulges from the beans inside.

NOTE: *There are bush beans and climbing, or pole, beans. Bush beans reach about 2 feet in height. Climbing types will grow up to 20 feet or more and need a trellis or other support system.*

Recommended Cultivars:

'Burpee's Tenderpod' (bush bean) produces exceptionally tender and delicious crisp pods. Adapts well to various growing conditions. Matures: 50 days.

'FORDHOOK No. 242' (bush lima) is a heavy yielder that produces 4–inch pods with three to four beans per pod. A terrific heat- and disease-tolerant variety. Matures: 75 days.

NOTE: *Bush beans mature faster than pole beans. However, pole beans will bear more and yield more bean pods over a longer time.*

TIP:

For climbing or pole type beans, set up a trellis or other support system before seeds are sown. The structures should be at least 6 feet tall. Sow seeds along the trellis, on both sides of the trellis. Once the plants have been thinned to about 6 inches apart, train them to grow up onto the trellis or netting. Another handy system is to erect a tepee out of bamboo or strong sticks or limbs. Sow seeds at the base of each stick or pole. Tie twine or nylon netting around the tepee to allow plants to clamber.

'Kentucky Wonder' (pole bean) is a classic pole known for its wonderful flavor and large 9–inch green pods. It can be used fresh, frozen, or dried for shell beans. Matures: 65 days.

'Roma II' (bush bean) is a bush form of the pole bean cultivar 'Romano' which bears prolific amounts of robust-tasting pods. It is very disease resistant. Matures: 53 days.

BEET, GARDEN

Beta vulgaris

Sowing Directions: Sow seeds directly outdoors 2 weeks before the last frost. Space seeds 2–4 inches apart within the rows, and space the rows 2 feet apart. Plant seeds ¼ inch deep. Make sure seeds are in good contact with the soil. Thin seedlings to 2–4 inches apart within the rows. Sow in late summer for a full crop.

'Burpee's Golden' beet

Days to Germination: 16–19

Garden Needs: Sow seeds in light, sandy, fertile, slightly acidic soil. Each "seed" actually contains up to 3–4 seeds, so careful thinning is important if the remaining seedlings are to have enough room to develop.

Days to Maturity: 50–80

Harvesting Tips: Beets are best harvested when they are a bit larger than a golf ball, though this will vary by cultivar. Beets can be pickled, canned, or used fresh for salads and soups.

Recommended Cultivars:

'Burpee's Golden' is a golden-colored beet with excellent sweetness and texture. Matures 55 days.

'Burpee's Red Ball' is a 3-inch-diameter beet for canning, pickling, or eating fresh. Matures: 60 days.

BROCCOLI

Brassica oleracea

Sowing Directions: This is a cool-weather long-maturing crop. Sow seeds indoors 6–8 weeks before the last frost into individual peat pots. About 2–3 weeks before the last frost, transplant the seedlings into the garden. Space plants 2 feet apart within the rows, and space the rows 2–3 feet apart. In southern zones,

'Green Goliath' broccoli

seeds can be sown directly into the garden in early fall. Sow seeds ¼–½ inch deep and space them 6 inches apart within the rows. Thin seedlings to 2–3 feet apart within the rows.

Indoor Germination Temperature: 70°F

Days to Germination: 10

Growing On Temperature: 50–60°F

Garden Needs: This vegetable needs cool, moist conditions. Plant in full sun in moist, rich, well-drained soil. Mulch the plants to keep the soil cool and moist.

Days to Maturity: 50–60, after transplanting to the garden

Harvesting Tips: Harvest when the heads form large, tight clusters and before the bud clusters begin to break apart and flower. Once the main head is harvested, smaller edible floret clusters will continue to be produced along the stem. Broccoli freezes well.

Recommended Cultivars:

'Bonanza Harvest' produces large (9-inch) heads and abundant and large shoots after the main head is harvested. Matures: 55 days, after transplanting to the garden.

'Green Goliath' is great for extending the harvest period of the main heads over a 3-week period. Matures: 55 days, after transplanting to the garden.

BRUSSELS SPROUTS

Brassica oleracea

Sowing Directions: This is most often grown as a fall crop in north-ern gardens and a fall and winter crop in frost-free zones. Seeds can be directly sown in the garden about 4 months before the first fall frost. Space seeds 12 inches apart within the rows, and space the rows 3 feet apart. Thin seedlings to 2 feet apart within the rows.

Indoor Germination Temperature: 70°F

Days to Germination: 10

Growing On Temperature: 50–60°F

Garden Needs: Plant in full sun in moist, fertile, well-drained soil. If the soil is poor, fertilize the seedlings about 2 weeks after planting out and again in about 4 weeks. Mulch plants to keep the soil moist and cool. They need 1–1½ inches of water each week.

Days to Maturity: 75–95

Harvesting Tips: Harvest the sprouts by cutting them off the main stem when they are the size of Ping-Pong balls and are firm. The buds growing lowest on the stem will mature first. You don't need to pick them if a light frost is predicted, as this will actually improve the flavor.

Recommended Cultivar: 'Jade Cross E' hybrid produces firm, closely set sprouts. Matures: 95 days.

CABBAGE

Brassica oleracea

Sowing Directions: This is a cool-weather crop that should be grown in spring or fall. For a spring crop, sow seeds indoors 5–7 weeks before the last frost into individual peat pots. Cover the seeds with ¼ inch of medium. Make sure the seedlings get at least 14 hours of light per day and cool growing on temperatures. Transplant the seedlings into the garden 2–3 weeks before the last frost, spacing them 2 feet apart within the rows, and spacing the rows at least 2½ feet apart. For a fall crop, sow seeds directly outdoors in midsummer in northern regions and in late summer in warmer regions.

Indoor Germination Temperature: 70°F

Days to Germination: 10–14

Growing On Temperature: 50–60°F

Garden Needs: Grow in full sun in deeply prepared, moist, fertile, well-drained soil. They like fertile conditions, so feed them with a complete fertilizer every 3–4 weeks. Mulch to keep the soil moist and cool. Keep them well watered.

Days to Maturity: 40–90, after transplanting to the garden

Harvesting Tips: Harvest when the heads are well rounded and firm. Cabbage can be used raw or cooked.

Recommended Cultivars:

'Earliana' (green) produces 5-inch round heads. Matures: 60 days, after transplanting to the garden.

'Fast Ball' (green) is a fast-growing variety that produces sweet and crunchy softball-size heads. Recommended for gardens with limited space. Matures: 45 days, after transplanting to the garden.

NOTE: *There are four types of popular cabbages: green, red, savoy, and Chinese. Green cabbage has smooth green leaves, red has red to purplish leaves, and savoy has dark green wrinkled leaves. Chinese cabbage (B. rapa) has an oblong shape with looser-growing, crisp, sweet leaves that have a light mustard or peppery flavor.*

'Orient Express' (Chinese) is a fast-growing, heat-resistant variety that produces small, blanched, crisp leaves with a peppery flavor. Matures: 43 days, after transplanting to the garden.

'Fast Ball' cabbage

CARROT

Daucus carota var. *sativus*

Sowing Directions: Before sowing, moisten the soil, since these tiny seeds can be displaced by watering. Sow seeds directly outdoors in early spring when the soil can be worked. Sow them evenly all along the rows and very lightly cover the

TIP:
If there is no space in your garden until other crops like spinach, peas, or lettuces mature, sow seeds indoors into individual peat pots 4 months before the first fall frost. Transplant the seedlings when other crops have been harvested or 4–5 weeks after germination.

Hybrid carrots are available in a range of sizes from golf ball to rocket size. Shorter cultivars are a good choice for gardens with heavy soils.

A   B   C   D   E   F

2"
3"
4"
5"
6"
7"
8"
9"

Choose the variety that's best for you: **A.** Orange Rocket **B.** Toudo **C.** Royal Chantenay **D.** Red Cored Chantenay **E.** Short'n'Sweet **F.** Thumbelina

TIP:

If your soil is heavy clay or otherwise difficult to prepare deeply, select a short carrot variety that grows only 3–4 inches in length like 'Short 'n Sweet' or 'Thumbelina,' which is a small, round variety.

seeds by filling a sieve with soil and evenly sifting it over them. Space rows 6–12 inches apart; when the seedlings emerge, thin them to 2–3 inches apart within the rows. Label the rows, since seeds can take up to 3 weeks to germinate. Can be sown in succession, every 3 weeks until late spring to early summer, to produce several crops into the fall. In fact, a light frost in the fall helps improve the flavor.

Days to Germination: 14–21

Garden Needs: Sow in full sun in loose, sandy, fertile, well-drained soil. Before sowing seeds, turn over and work the soil to a depth of 14 inches or more. Remove rocks and other debris from the soil. If the soil is a heavy clay, sow the seeds into raised beds. In either case, incorporate lots of well-aged manure and other organic matter. Add some granular fertilizer to the planting area before the seeds are sown. Fertilize the seedlings when they are 6 inches tall. Make sure the plants and soil are watered regularly.

Days to Maturity: 65–75

Harvesting Tips: To see if they are ready to be dug up, brush back the soil from below the foliage area.

Carrot tops should be 1–2 inches in diameter before picking. Depending on the variety, some carrots can be harvested smaller for use in salads and for snacks. Harvest the roots by using a shovel, spade, or pitchfork and gently digging them up.

Recommended Cultivars:

'Nantes Half Long' produces 7-inch-long roots that are especially delicious. Matures: 70 days.

'Toudo' hybrid' develops quickly so these sweet roots can be harvested when very small (3–4 inches) or left to grow to maturity (7½ inches). Matures: 70 days.

CAULIFLOWER

Brassica oleracea

Sowing Directions: This is a cool-weather crop. For best results, sow seeds indoors 8 weeks before the last frost. Cover seeds with ¼ inch of medium. When seedlings are 2 inches tall, transplant into individual containers. These seedlings can be transplanted out 2–3 weeks before the last frost. Space about 3 feet apart within the rows, and space the rows 3 feet apart. For a second crop, sow seeds directly into the garden 3–3½ months before the first fall frost. Sow seeds to a depth of ¼ inch and space seeds 1 foot apart within the rows. Thin seedlings to 3 feet apart.

Indoor Germination Temperature: 70–75°F

Days to Germination: 10–14

Growing On Temperature: 50–60°F

Garden Needs: To produce a good crop, this vegetable needs uniform growing conditions—ideally, full sun; cool weather; plenty of even moisture; and well-prepared, rich, well-drained soil. Mulch to keep the soil moist and cool. Fertilize the seedlings when they are a few inches tall and then again about 4 weeks later.

Days to Maturity: 45–70

Harvesting Tips: Harvest when heads have enlarged but the florets within the head are still tight. If you want to blanch them (to keep the heads white), take foliage and wrap it around the flower head while it is still small—about 2–3 inches in diameter and keep the foliage tied around the head until it matures. There are self-blanching varieties as well as green and purple cultivars that do not need blanching.

Recommended Cultivar: 'Early White' hybrid is an early-maturing variety that produces large (9-inch), white, round heads. It is an excellent variety for freezing. Matures: 52 days.

'Early White' hybrid cauliflower

CELERY

Apium graveolens var. *dulce*

Sowing Directions: This is a cool-climate crop. Sow seeds indoors 10–12 weeks before the last frost into individual growing containers. The plants need a lengthy growing season. Space seedlings 6–10 inches apart within the rows, and space the rows 2–3 feet apart.

Indoor Germination Temperature: 70–75°F

Days to Germination: 21–25

Growing On Temperature: 60°F

Garden Needs: Plant in full sun in deeply dug and well-prepared, light, rich, moist to wet soil. The vegetable needs a long growing season, so it is best to plant as soon as the soil can be worked. Keep the soil very moist throughout the growing season and fertilize the plants every 2 weeks with a complete fertilizer.

Days to Maturity: 80–140

Harvesting Tips: Harvest stalks by cutting the base at soil level with a knife.

Recommended Cultivar: 'Tall Utah 52–70 R Improved' produces super-crunchy stalks. Matures: 105 days.

COLLARD AND KALE

Brassica oleracea

Sowing Directions: To get a jump on the season, sow seeds indoors in late winter and transplant the seedlings into the garden in early spring. For best results, sow seeds ¼ inch deep directly outdoors in early spring as soon as soil can be worked. Space the seeds 4–6 inches apart within the rows, and space the rows 2–3 feet apart. Thin seedlings so remaining seedlings are 2 feet apart. A second crop can be grown in late summer or early fall. Collard tolerates heat better than kale, so it can also be sown later in the spring for summer growing on. Kale should be grown as a spring or early fall crop.

Indoor Germination Temperature: 70°F

Days to Germination: 7–10

Growing On Temperature: 50–60°F

Garden Needs: Plant in full sun in fertile, loose, well-drained soil. If the soil is poor, fertilize the seedlings when they reach a few inches in height. Mulch and keep the soil evenly moist.

Days to Maturity: 60–70

Harvesting Tips: Leaves are ready to harvest when they are 8–10 inches long. Pick the outer leaves and allow the younger leaves (growing from the center) to mature. Leaves can be used fresh or steamed. A light frost improves the flavor of these vegetables.

Recommended Cultivars:

'Dwarf Blue Curled Vates' (kale) is a very cold-tolerant variety. Matures: 55 days.

'Georgia' (collard) tolerates summer heat and cold temperatures very well. Matures: 60 days.

CORN

Zea mays

Sowing Directions: Soak seeds before sowing, especially the super-

TIP:
Cauliflower is easy to store by freezing. Break the heads into floret pieces, put them in a zipper-locking plastic bag, and freeze.

sweet types with very shriveled seeds. Sow seeds directly into the garden after the last frost when soil has warmed. Do not sow into cold and wet soils or the seeds may rot before germinating. Cover seeds with 2 inches of soil. Sow seeds 4–6 inches apart within the rows and space the rows 2–3 feet apart. Thin the seedlings to 12 inches apart within the rows. In areas with short growing seasons, sow seeds indoors a few weeks before the last frost into individual peat pots. Transplant them into the garden when the seedlings are about 4 inches tall.

Indoor Germination Temperature: 70–75°F

Days to Germination: 5–7

Growing On Temperature: 60°F

Garden Needs: Plant in full sun in deeply dug, fertile, loose, well-drained, slightly acidic soil (optimal pH is 5.8–6.5). Incorporate generous amounts of well-aged compost and manure. These plants are heavy nitrogen users; side dress with a complete fertilizer high in nitrogen when they reach 6 inches in height and fertilize again in 3–4 weeks. Be

NOTE: *Corn is wind pollinated. It is important to have more than one row, even if you have only a half dozen plants. Furthermore, it is important to isolate corn varieties to avoid hybridization problems caused by wind pollination. Separate varieties by at least 250 feet or make sure that the varieties mature at least 14 days apart.*

sure to keep the plants well watered (from tassel development through to harvest); they need at least 2 inches of water per week throughout the growing season.

Days to Maturity: 63–92

Harvesting Tips: Harvest when the tassels turn brown and dry. Or pull back the husks from an ear and use a fingernail to test if the kernels are tender, succulent, and produce a milky sap when pricked. If the sap is watery, the corn is not yet ready; if it is creamy, the corn is overripe.

Recommended Cultivars:

'Breeder's Choice' produces extra-sweet, creamy textured, yellow kernels. Matures: 73 days.

'Early Choice' is a golden kernel variety that is good for short seasons. Matures: 66 days.

'Breeder's Choice' corn

TIP:
To help ensure good germination rates (especially in northern gardens), cover the rows with black plastic a few weeks before the last frost to help warm up the soil before sowing seeds. The plastic can be removed or kept in place throughout the growing season. If the plastic is kept, cut out growing holes. Also, seeds can be treated with a fungicide to prevent disease problems.

TIP:
Cucumbers can also be sown into rows. Space the seeds 1 inch apart within the rows, and space the rows 3–4 inches apart.

'Illini Xtra-Sweet' is a delicious supersweet yellow variety that stays sweet tasting after picking. Matures: 85 days.

'Silver Queen' is a corn with a cult following! A white kernel variety renowned for its sweet and tender taste and large (9-inch) ears. Matures: 92 days.

CUCUMBER

Cucumis sativus

Sowing Directions: For indoor sowing, sow seeds 4–6 weeks before the last frost into individual peat pots that can be transplanted directly into the garden. Cover the seeds with 1 inch of medium. Thin seedlings to two per pot. After the last frost, plant the entire pot and do not try to separate the seedlings. For outdoor sowing, after the last frost create small mounds of soil about 4 inches in height. Plant four to five seeds evenly around the mound and cover them with 1 inch of soil. Space the mounds about 2 feet apart within the rows, and space the rows 3–4 feet apart. Thin seedlings so there are two to three per mound.

Indoor Germination Temperature: 70°F

Days to Germination: 7–10

Growing On Temperature: 60–65°F

Garden Needs: Plant in full sun in loose, fertile, well-drained, slightly acidic soil. When the seedlings are a few inches tall, apply fertilizer at one-quarter strength and fertilize again every 4 weeks. Apply mulch to keep the soil moist and cool. If space is limited, the plants can be trained to grow up a fence or trellis. Some dwarf varieties are good for growing in trellised containers.

Days to Maturity: 48–70

A-frame cucumber trellis

Harvesting Tips: Harvest when the fruit has turned dark green and has reached the average size for the variety. Most cucumbers grow to 6–9 inches in length. Regularly pick ripe cucumbers as they mature, usually over a 5–6-week period. If ripened fruit is left on the vine and begins to yellow, it will delay or stop further fruit development. For pickling, harvest when smaller, 2–3 inches in length, or grow a pickling variety.

NOTE: *There are vine-type and bush-type cucumbers. Vining types should be grown with some trellis or other support. Bush cucumbers can be left to grow on the ground.*

Recommended Cultivars:

'Burpee Pickler' produces medium green fruit that is excellent for picking when young for pickling, and is good for eating when more mature. Matures: 53 days.

'Bush Champion' is one of the best compact bush plants with good disease-resistance. Matures: 55 days.

'Early Pride' hybrid produces amazing amounts of dark green 8½-inch fruit over many weeks. A good disease-resistant variety. Matures: 55 days.

'Sweet Success' produces 14-inch, dark green fruit with delicious, crisp flesh. A very disease-resistant variety. Matures: 58 days.

EGGPLANT

Solanum melongena

Sowing Directions: Sow seeds indoors 6–8 weeks before the last frost into individual peat pots that can be directly transplanted into the garden. Sow seeds ¼ inch deep. Space seedlings 2–3 feet apart within the rows, and space the rows 3–4 feet apart.

Indoor Germination Temperature: 75–80°F

Days to Germination: 7–14

Growing On Temperature: 70°F

Garden Needs: After the last frost and when the soil has warmed up a bit, plant in rich, deeply worked, well-drained soil. To ensure warm soil, lay black plastic over the rows to warm the soil for a few weeks before planting. The plastic can be left for the entire growing season to keep soil warm and moist. Fertilize seedlings with a light, complete fertilizer after transplanting and then every 4 weeks to maturity. Keep plants consistently well watered .

Days to Maturity: 54–80

Harvesting Tips: Harvest the fruit when it begins to reach mature size, which varies by cultivar. Don't allow the fruit to turn to brownish before picking; harvest fruit immediately as it matures, as this will encourage more fruit production.

Recommended Cultivars:

'Burpee' hybrid is a very productive variety with medium, shiny purple fruit. Matures: 70 days, after transplanting in the garden.

'Millionaire' hybrid has early maturing fruits with a slender (2 inches), cylindrical, long (12 inches) shape. A Japanese type with delicious flavor. Matures: 55 days, after transplanting to the garden.

ENDIVE

Cichorium endivia

Sowing Directions: Sow seeds directly into the garden in early spring 2–3 weeks before the last frost. Seeds need light to germinate; cover very lightly with just ⅛ inch of soil. Space seeds 4 inches apart within the rows, and space the rows 1½ feet apart. Thin the seedlings so remaining seedlings are spaced 12 inches apart within the rows. For a fall crop, sow seeds directly outdoors in mid or late summer (or early fall in warmer climates).

Days to Germination: 7–10

Garden Needs: Plant in full sun in fertile, well-drained soil with lots of organic matter. They prefer cool growing conditions. Keep the plants consistently well watered. The plants can withstand a few light frosts.

Days to Maturity: 85–90

Harvesting Tips: The inner center or heart of the vegetable is the most tender and delicious. Blanching improves the flavor by making it less bitter. To blanch the endive, tie the up outer foliage around the center leaves and secure it with a piece of nylon or twine. Do this when the foliage is dry to avoid any rot problems. After 2–3 weeks of blanching, the heads are ready to be harvested. The hearts should be whitish or light yellow.

Recommended Cultivar: 'Green Curled Endive' has dark green, finely cut 18-inch heads. Matures: 90 days.

KOHLRABI

Brassica oleracea

Sowing Directions: This is a cool-weather plant. Sow seed indoors 4–5 weeks before the last frost. Or sow seeds directly outdoors in early spring as soon as the soil can be worked. Cover seeds with ½ inch of medium. Space seeds 6 inches apart within the rows, and space the rows 1–1½ feet apart. Thin seedlings to 6 inches apart within the rows. In southern zones, a fall crop can be directly sown into the garden in late summer.

TIP:
To prepare this vegetable, peel the outer layer of skin and discard. Eat kohlrabi raw or cooked. The leafy tops can be used raw in salads or eaten steamed.

Indoor Germination Temperature: 70°F

Days to Germination: 10–14

Growing On Temperature: 50–60°F

Garden Needs: Plant in full sun to light shade in rich, moist, well-drained, slightly acidic soil. Keep the plants evenly moist, which can be accomplished with a good mulching. If the soil is poor, fertilize the seedlings when they're transplanted and again in about 4 weeks.

Days to Maturity: 45–55

Harvesting Tips: For best taste, harvest the bulbs when they reach 2–3 inches across. Cut the bulbs at the rounded, swollen base of the stem.

'Grand Duke' hybrid kohlrabi

Recommended Cultivar: 'Grand Duke' hybrid is a disease-resistant All-America winner that matures early and produces crisp, white bulbs with a mild, sweet flavor. Matures: 45–50 days.

LEEK. SEE ONION

LETTUCE

Lactuca sativa

Sowing Directions: First moisten the soil, since these tiny seeds can be displaced by watering. Sow seeds directly outdoors in early spring when soil can be worked. Sow evenly all along the row, spacing the rows 2 feet apart. Seeds need light to germinate; fill a sieve with soil and evenly and lightly sift it over the seeds. Thin seedlings to 4 inches apart for leaf lettuce and 12 inches apart for head lettuce. Another sowing can be made in late summer or early fall in warmer climates for a fall crop. If growing in cooler regions, seeds can be sown into a cold frame in early fall.

Days to Germination: 7–10

Growing Needs: When the weather is still cool, plant in full sun or part shade in rich, loose, well-drained, moist soil with lots of organic matter worked into it. Add a complete fertilizer to the soil before planting. If the soil is poor, use a liquid fertilizer when the plants are about 4 inches tall. It's very important to keep the soil evenly well watered.

Days to Maturity: 45– 75

NOTE: *There are two main types of lettuce: leaf and head. Leaf (or looseleaf) lettuces produce a number of upright growing leaves, and head lettuces form a central, round, tighter bunch of leaves. Lettuce is also available in different leaf colors and leaf types.*

Harvesting Tips: Harvest by carefully pulling the lettuce up from the base of the plants or by cutting it just below the base with a sharp knife. Lettuce must be harvested before it bolts (produces a flower stalk), because after that point, the leaves become bitter. Plants will bolt when temperatures begin to get above 80°F.

Recommended Cultivars:

'Buttercrunch' (butterhead) has compact heads with a delicious buttery smooth texture. Recommended for cooler growing areas. Matures: 75 days

'Green Ice' (looseleaf) produces sweet, dark green leaves and is slow to bolt. Matures: 45 days.

'Summertime' (head) has a sweet flavor and excellent crisp texture. A heat-tolerant variety. Matures: 70 days.

MELON: MUSKMELON, CANTALOUPE, AND HONEYDEW

Cucumis melo

Sowing Directions: In northern zones, sow seeds indoors 4–6 weeks before the last frost into individual peat pots that can be transplanted directly into the garden, to avoid transplant shock. Cover seeds with ½ inch of medium. Transplant seedlings into the garden after all danger of frost has passed and the soil has warmed up a bit. Space seedlings 3–4 feet apart within the rows, and space the rows 5–6 feet apart. (Or plant in mounds, as directed for southern zones.) Space rows 5–6 feet apart. In southern zones, sow seeds directly into the garden after the last frost and once the soil has

warmed up a bit. Sow four to five seeds ½ inch deep into mounds or hills of soil. Space the mounds 4 feet apart within the rows, and space the rows 5–6 feet apart. Thin seedlings to two to three per mound to grow on.

Indoor Germination Temperature: 75–80°F

Days to Germination: 6–10

Growing On Temperature: 65–70°F

Garden Needs: Plant in full sun in light, fertile, well-drained soil. If the drainage needs to be improved, add builder's sand. It is important to keep the plants evenly well watered. Taper off watering when the fruit is in the final stages of ripening. Mulch around plants with black plastic or black landscape fabric to keep soil temperatures warm and the soil moist. Feed with a complete light fertilizer every 3–4 weeks. If space is a problem, grow the plants up a fence or trellis; tie up the vines with a nonbinding material like nylon or strips of cloth.

Days to Maturity: Muskmelon and cantaloupe, 72–90; honeydew, 90–110

Harvesting Tips: Melons are ripe when their stems easily twist off the vines. Honeydew turns a light greenish yellow when ripe. You can also smell melons for ripeness.

Recommended Cultivars:

'Burpee' hybrid (cantaloupe) is beloved by gardeners for its superior flavor and thick, firm, juicy, deep orange flesh. The fruit is ribbed and netted and grows to a

NOTE: *Some varieties of melon, especially honeydews, need a long, hot growing season.*

mature weight of about 4½ pounds. Matures: 82 days.

'Burpee's Ambrosia' hybrid (cantaloupe) is a luxurious-tasting melon with incredibly sweet and juicy peach flesh. A very prolific variety that produces extra-delicious fruit that matures at about 5 pounds. Resistant to powdery mildew. Matures: 86 days.

'Venus' hybrid (honeydew) is an early maturing variety that produces 5 by 6-inch oval fruit with an excellent juicy sweetness and a luscious aroma. Matures: 88 to 90 days.

'Venus' hybrid honeydew melon

MUSTARD

Brassica juncea var. *crispifolia*

Sowing Directions: Sow seeds directly into the garden in early spring about 2–3 weeks before the last frost. Cover seed with ¼–½ inch of soil. Space seeds 2–3 inches apart within the rows, and space the rows 1 foot apart. Thin seedlings to 8–10 inches apart within the rows. For a second crop, sow seeds 2–4 weeks after the initial sowing. In southern climates, sow again in the fall for a winter crop.

Days to Germination: 8–12

Garden Needs: Plant in rich, loose, well-drained soil with a lot of organic matter—compost and well-aged manure—mixed into it. If the soil is poor, fertilize after the seedlings emerge. To keep the soil moist and cool, mulch with a light compost. Keep plants well watered.

Days to Maturity: 35–40

Harvesting Tips: The younger green leaves are the most tender and delicious. Plants grown in hot weather can have a bitter taste, so make sure to grow it during the cooler months. Use mustard fresh in salads or steam it. If growing it in the fall, pick leaves after a light frost, since a frost improves the flavor.

OKRA

Abelmoschus esculentus

Sowing Directions: Soak seeds in warm water for 24 hours to help speed germination. Sow seeds indoors into individual peat pots 4–6 weeks before the last frost in all growing areas except Zones 9–11. In those Zones, sow seeds directly in the spring. In other zones, seed can be sown directly outdoors after soil has warmed up—after the last frost date, or when nighttime temperatures remain above 55°F. Whether started indoors or outdoors cover seeds by ½ inch with medium. Space or thin seedlings to 2 feet apart within the rows, and space the rows 3 feet apart.

Indoor Germination Temperature: 70–75°F

Days to Germination: 10–14

Growing On Temperature: 60–65°F

Garden Needs: After nighttime temperatures remain above 50–55°F, plant in full sun in fertile, well-drained soil. Provide a light monthly fertilizer feeding.

Days to Maturity: 48–60

Harvesting Tips: Harvest most varieties when the pods are young (3–4 inches long), every 2 or 3 days to extend the harvest. Pod sizes vary by cultivar, ranging from 3 to 6 inches long. Okra is delicious sautéed or stewed with tomatoes. It's used in soups and it's a classic ingredient for gumbos, stews, and soups.

'Clemson Spineless' okra

TIP:
Mustard is a cool-weather crop. So summer growing in all but the coolest climates is not recommended.

TIP:
Many gardeners start onions from sets or plants. Onions require a long growing season, and some varieties can take up to 300 days to reach maturity.

Recommended Cultivars:

'Clemson Spineless' is an All-America winner that produces tasty dark green pods. Matures: 56 days.

'North & South' is an excellent choice for northern regions with cooler growing seasons. Matures: 48–50 days.

ONION (SCALLION) AND LEEK

Allium cepa: onion (scallion), *A. ampeloprasum*: leek

Sowing Directions: Sow seeds indoors 8–10 weeks before the last frost. Seed can be sown outdoors in early to midspring after the soil can be worked. Cover seeds by ¼ inch with medium. Space seeds outdoors 2–4 inches apart within the rows, and space the rows 2–3 feet apart. Thin seedlings to 4–6 inches within the rows.

Indoor Germination Temperature: 70°F

Days to Germination: 7–14

Growing On Temperature: 60°F

Garden Needs: Plant in full sun in rich, well-drained soil. Make sure the soil has been worked to a depth of at least 12 inches and that generous amounts of compost have been mixed in so the vegetables have a light, loose soil to develop in. Keep the plants well watered and fertilize them early in the growing season. Transplant leek seedlings into 4–6-inch deep trenches. As leeks grow, soil should be gently hilled up against the stems just under where the foliage begins to keep the stems white.

Days to Maturity: 95–300

Harvesting Tips: Onions are ready to be pulled up when the foliage begins to yellow and fall over. Allow to fully dry before storing in a cool, dry location. Scallions are green onions that have not formed bulbs; grow a variety specifically recommended for early harvest. Leeks are harvested in the fall.

Recommended Cultivars:

'Burpee Sweet Spanish' hybrid (onion) produces large, round onions with yellow skin and white, mild flesh. Stores very well. Matures: 110 days.

'Evergreen Long White Bunching' (scallion) produces long, slender stalks. Matures: 60–120 days.

'Sweet Sandwich' hybrid (onion) is an extra-sweet variety with light yellow flesh and light brown skin. Delicious raw or cooked. Keeps well. Matures: 105 days.

'Titan' (leek) is an early maturing variety with large (8-inch-long) stalks that are 2 inches wide. Matures: 110 days.

PARSNIP

Pastinaca sativa

Sowing Directions: Before sowing, soak seeds overnight in warm water. Sow seeds directly outdoors in early spring as soon as soil can be worked. Cover the seeds with ½ inch of soil.

Sow seeds evenly in rows, and space the rows 1–2 feet apart. Be sure to label the rows since seedlings can take more than 3 weeks to germinate. Thin the seedlings to 4–6 inches apart. In Zone 8 and warmer, seeds can also be sown again in early fall for an early spring crop the following year.

Days to Germination: 21–27

'Hollow Crown' parsnip

Garden Needs: Plant in full sun in deeply prepared, rich, loose, well-drained soil that contains lots of organic matter. Work the soil to a depth of at least 14 inches and remove stones, rocks, and other debris from the soil before sowing seeds. Keep the plants well watered especially for at least 4 weeks after the seedlings begin to emerge.

Days to Maturity: 105

Harvesting Tips: The roots can be dug up in early to late fall. Cooler weather and a light frost will improve the flavor.

Recommended Cultivar: 'Hollow Crown' produces 12-inch-long roots that have a mild, white flesh. A light frost improves the sweet, nutty flavor. Matures: 105 days.

PEA, SNOW PEA, SUGAR PEA

Pisum sativum

Sowing Directions: This is a cool-weather crop and shouldn't be sown after temperatures remain above 75–80°F. Sow seeds directly outdoors in early spring as soon as soil can be worked. These will be the first seeds in the garden. Plant seeds 2 inches deep. Space seeds 6 inches apart within the rows, and space the rows 2 feet apart. Thin the seedlings to 1 foot apart (dwarfs) or 2 feet apart (others), depending on variety. Sow seeds successively every 2 weeks until late spring and again in mid to late summer for a fall crop.

Days to Germination: 7–10

Garden Needs: Plant in full sun in rich, sandy, deeply prepared, well-drained soil. When seedlings are a few inches tall, fertilize with a light, complete fertilizer and fertilize again in 3–4 weeks. Vining types will need a trellis or other support system.

Days to Maturity: snow and sugar, 65; English, 56–70; snap, 64–70

Harvesting Tips: Snow and sugar peas should be harvested when still flat and tender. Both pods and peas are edible. English peas should be harvested when the pods swell yet are still succulent and soft. Shell these peas before eating. Snap peas

TIP:
Mulch leeks in the fall to extend the harvest season. A light frost improves their sweet, succulent taste.

TIP: Another way to plant peas is to sow two rows spaced 4–6 inches apart. Sow seeds 4 inches apart within the rows, and space the double rows 2 feet apart.

TIP: Wrinkled pea varieties are considered sweeter tasting than smooth peas.

should be harvested when swollen yet on the young side, for best taste. Both pods and peas are edible.

'Maestro' English pea

Recommended Cultivars:

'Maestro' (English) is an especially heavy producer with pods that yield 9–12 medium-size peas. Matures: 61 days.

'Oregon Sugar Pod II' (snow) is a vigorous-growing variety that produces a huge crop. It is resistant to pea virus, wilt, and powdery mildew. Matures: 68 days.

'Super Snappy' (snap) produces the largest edible-podded pea available. Pods are sweet and crisp. It is very disease-resistant. Matures: 65 days.

PEANUT

Arachis hypogaea

Sowing Directions: Shell the seeds before sowing. Sow seeds directly outdoors after the last frost. Cover them with about 1 inch of soil. Space sseeds 3 feet apart within the rows and space the rows 3 feet apart.

NOTE: *Peanuts need a long, hot growing season to fully mature by frost. Do not plant peanuts in areas that do not have at least 4–5 months of frost-free weather.*

Days to Germination: 18–25

Garden Needs: Plant in light, well-drained, deeply dug and prepared, alkaline soil. Keep plants watered during dry periods.

Days to Maturity: 110–125

Harvesting Tips: Peanuts are ready to harvest in late summer or early fall when the foliage yellows and wilts. Shake off the soil from the pods and allow them to fully dry indoors for 2–4 weeks before shelling and eating.

Recommended Cultivar: 'Jumbo Virginia' is a very tasty variety; with good growing conditions, one plant will yield 50–60 pods. Matures: 120 days.

PEPPER

Capsicum annuum var. *annuum*

Sowing Directions: This is a warm-weather crop that needs a long growing season. Outdoor sowing is not recommended. Sow seeds indoors 8–10 weeks before the last frost. Cover seeds with ¼ inch of medium. Space seedlings 2 feet apart within the rows, and space the rows 3 feet apart.

Indoor Germination Temperature: 75–80°F

Days to Germination: 10

Growing On Temperature: 65°F

Garden Needs: In late spring or early summer, transplant the seedlings when the soil has warmed up and nighttime temperatures remain above 55°F. Plant in fertile, well-drained soil. If the soil is poor, fertilize the seedlings lightly at planting time and again in 4–5 weeks. Keep them well watered, especially during fruit set and development.

Days to Maturity: 60–95, after transplanting to the garden

Harvesting Tips: From the sweet bell peppers to the hot habanero, most peppers will turn a bright or dark brown red when they are fully ripened. Note, however, that there are several yellow, purple, and orange varieties. Some peppers, like bell peppers and cayennes, can be picked green just before they turn red and after they've reached their mature size. Peppers are wonderful raw, stir-fried, grilled, roasted, stuffed, and more.

Recommended Cultivars:

'Crispy' hybrid (sweet) produces exceptionally heavy yields of crispy, thick-walled, 3½ by 2¾-inch peppers. Matures: 70 days, after transplanting to the garden.

'Little Dipper' hybrid (sweet) bears prolific quantities of miniature peppers that are terrific for salads and dipping. Matures: 66 days, after transplanting to the garden.

'Thai Dragon' hybrid (very hot) bears many narrow, 3½-inch-long peppers, which are easily dried. For the hot pepper lover, this one is five times hotter than the jalapeño. Matures: 70 days, after transplanting to the garden.

'Zippy' hybrid (mildly hot) produces a thin, hot but not fiery, 6-inch-long pepper. Matures: 57 days, after transplanting to the garden.

PUMPKIN

Cucurbita maxima

Sowing Directions: This crop needs a long, hot growing season. Sow seeds indoors 4–6 weeks before the last frost in individual peat pots that can be transplanted directly into the garden. Plant three to four seeds per container and cover them with 1 inch of medium. Thin to two seedlings per pot. Can be

'Zippy' hybrid sweet bell pepper

sown directly into the garden after the last frost. Create mounds or hills of soil within rows and sow four to five seeds per mound. Space the mounds 3–4 feet apart within the rows, and space the rows 6 feet apart. Thin to two seedlings per mound.

Indoor Germination Temperature: 70–75°F

Days to Germination: 7–10

Growing On Temperature: 60–65°F

Garden Needs: Plant in full sun in light, sandy, well-drained, fertile soil with plenty of organic matter mixed in. When the seedlings are 6 inches tall, give them a complete fertilizer application and then fertilize again every 4 weeks. Keep the plants evenly well watered.

Days to Maturity: 90–120

Harvesting Tips: Pumpkins can be harvested when they reach mature size (which varies by cultivar) and have a firm rind and their vines begin to yellow and die back. A light frost or two will not harm pumpkins. If you leave at least a few inches of the stem attached to the pumpkin, it will keep longer. Seeds are edible.

Recommended Cultivars:

'Ghost Rider' is an excellent dark orange carving or cooking pumpkin that grows to 16 inches and up to 20 pounds. Matures: 115 days.

'Bushkin' is a good choice for smaller garden areas. Plants spread only 6 feet but produce from one to three bright

TIP:
To make growing radishes super easy (especially for beginning gardeners), plant seed tapes. The seed is very fine and seed tapes make handling much easier and sowing more exact.

orange-colored 10-pound fruits each. Matures: 95 days.

'Prizewinner' hybrid pumpkin

'Triple Treat' is a great all-around pumpkin for carving, pies, and seeds. Bright orange fruits grow to 9 inches and weigh 8 pounds. Matures: 110 days.

Novelty Pumpkins:

'Jack Be Little' miniature orange pumpkins grow to just 3 inches across and 2 inches high. Matures: 95 days.

'Lumina' is an all-white pumpkin good for fall decoration or carving. Fruits are 8–10 inches across and weigh 10–12 pounds. Matures: 80–90 days.

RADISH

Raphanus sativus

Sowing Directions: This is a fast-growing, cool-weather crop. For best results, sow directly into the garden from 2 weeks before the last frost up to midspring, in 2-week intervals if successive harvests are desired. Cover seeds by ½ inch with medium. Sow seeds evenly down the rows, and space the rows 8–12 inches apart. Thin seedlings to 1–2 inches apart. For a fall crop, sow in late summer in northern regions and in early fall and winter in southern zones.

Days to Germination: 4–6

Garden Needs: Plant in full sun in sandy, loose, well-drained soil. Work the soil to a depth of at least 12 inches before sowing. When seedlings are a few inches tall, lightly fertilize with a complete balanced fertilizer. Keep the soil consistently moist throughout the growing season. If the plants appear to be heaving up from the soil, cover the roots with additional earth.

Days to Maturity: 22–45

Harvesting Tips: Most radishes are ready to harvest when they reach the size of cherry tomatoes. Other varieties, such as 'White Icicle', develop longer roots.

Recommended Cultivars:

'Cherry Belle' is an early-maturing, scarlet variety with crisp white flesh. Matures: 22 days.

'White Icicle' has a 5-inch-long cylindrical white root with a crisp flesh and mild pungency. Matures: 28 days.

'Summer Cross' hybrid is a daikon- or Oriental-type radish with 6-inch-long white roots that have a crisp and mild taste. Matures: 45 days.

RUTABAGA, SWEDISH TURNIP

Brassica napus

Sowing Directions: This is a cool-weather crop. For an early summer crop, sow seeds directly outdoors 3–4 weeks before the last frost. For a fall crop, sow seeds directly outdoors 12–14 weeks before the first fall frost (midsummer for most gardeners). In southern climates, sow seeds directly outdoors in early fall. Cover seeds with ½ inch of soil. Space seeds about 2 inches apart within rows and space the rows 1 foot apart. Thin seedlings to 6 inches apart within rows.

Days to Germination: 7–10

Garden Needs: Plant in rich, loose, well-drained, well-prepared soil with lots of organic matter. Keep soil consistently moist so the plants stay well watered.

Days to Maturity: 85–95

Harvesting Tips: Rutabagas taste best and sweetest when allowed to grow fairly large.

Recommended Cultivar: 'Burpee's Purple-Top Yellow' is a sweet-tasting variety. Matures: 90 days.

SPINACH

Spinacia oleracea

Sowing Directions: This is a cool-weather crop. Sow seeds directly into the garden in early spring as soon as the soil can be worked. Sow seeds evenly down rows and cover with ½ inch of soil. Space rows 1½–2 feet apart. Thin seedlings to 4–6 inches apart. If desired, seeds can be sown in 1–2-week intervals up until midspring. Another sowing can be made later in the growing season,

TIP: Seed tapes are an easy way to sow spinach seeds.

TIP: Mulch the planting rows with 3-foot-wide black plastic. Squashes thrive with the added warmth the plastic adds to the soil. If you use plastic, it's easiest to irrigate plants with a drip irrigation system or soaker hoses placed under the plastic. Cut a cross into the plastic with a knife, and plant the seedlings into the hole. Secure the plastic with soil, rocks, and/or bricks.

about 40 days before the first fall frost. It can be sown as a winter crop in frost-free zones.

Days to Germination: 8–12

Garden Needs: Plant in full sun in rich, fertile, very well drained, neutral soil. Be sure to keep the plants well watered and to incorporate generous amounts of compost and other organic matter into the soil.

NOTE: *Spinach seedlings are susceptible to damping off. Make sure the soil is very well drained and the seeds are kept moist but not overwatered.*

Days to Maturity: 42–50

Harvesting Tips: It is best to harvest an entire spinach plant before it bolts (develops a flower stalk). Younger leaves are good for using raw in salads. Older leaves are usually cooked.

'Melody' hybrid spinach

Recommended Cultivars:

'Avon' hybrid is a fast-growing, good-tasting variety that produces large, dark green, crinkled leaves. Matures: 44 days.

'Melody' hybrid is trusted for its disease-resistance and large, dark green, thick leaves. Matures: 42 days.

SQUASH (WINTER, SUMMER)

Cucurbita

Sowing Directions: Sow seeds indoors 4–6 weeks before the last frost in individual peat pots that can be directly transplanted into the garden; be sure to harden off seedlings before planting them outside. Seeds can be sown directly outdoors, after the last frost, into mounds or hills of soil about 1 inch deep. Space the hills 3 feet apart within the rows, and space the rows 3 feet apart for bush types and 6 feet apart for vining types.

Indoor Germination Temperature: 70°F

Days to Germination: 7–10

Growing On Temperature: 60–65°F

NOTE: *There are two types of squash growth habits: bush and vining. Bush plants have a compact, rounded habit. Vining plants grow horizontally along the ground. To save space, vining plants can be trained to grow up a trellis or other support. Bush plants are good for container gardens.*

Garden Needs: Plant in full sun in rich, fertile, well-drained soil with lots of incorporated organic matter. If the soil is poor, fertilize the seedlings when they are 4–6 inches tall and then fertilize again every 4 weeks. Keep the plants well watered, especially during fruit development.

Days to Maturity: Summer squash, 48–85; winter squash, 85–110

Harvesting Tips: Summer squash (crookneck, zucchini, patty pan, etc.) is best when harvested young. Unlike winter squash, summer squash does not have a hard shell when mature and matures much faster than winter squash. Harvest the plants often so they are encouraged to produce more fruits. Winter squash (acorn, buttercup, butternut, cushaw, spaghetti, Hubbard, etc.) should be harvested when fully mature, after the shells have hardened. For some winter squash, harvest after the vines have been killed by frost. Keep 1–4 inches of the stem attached to the fruit to help preserve it. After harvesting cure winter squash for a week in a warm location which will help preserve them. Store after curing in a cool basement or similar location.

Recommended Cultivars:

'Burpee Hybrid Zucchini' (summer, bush) produces green, tender, and very delicious fruit. It's exceptionally prolific. Matures: 50 days.

'Butterstick' hybrid (summer, vine) is a very prolific variety with golden yellow skin and a white, sweet, nutty-tasting flesh. Matures: 50 days.

'Butterstick' hybrid summer squash

'Lakota' (winter, vine) is an heirloom variety with a fine-grained, sweet, and nutty-tasting orange flesh. Matures: 85 days.

'Vegetable Spaghetti' (winter, vine) produces very delicious 2–3-pound fruit in late summer and fall. Interior contains long thin strands of flesh which resembles cooked spaghetti. Matures: 100 days.

STRAWBERRY

Perennial
Fragaria vesca
Zones: 4–8

Sowing Directions: For best results, choose a variety that will bear fruit the first year and sow seeds indoors in late winter about 10–12 weeks before the last frost. Transplant seedlings when they are 2 inches tall into larger growing on containers. Give plants a dilute, balanced fertilizer feeding every 2 weeks. Harden off the seedlings before planting them in the garden in early to midspring. Seeds can be sown directly into the garden in early fall or early spring as soon as the soil can be worked. Sow seeds in rows and cover them within ¼ inch of soil. Space seeds 6 inches apart within the rows, and space the rows 3 feet apart. Thin or plant seedlings so they are 1 foot apart for everbearing varieties and 2 feet apart for single-crop varieties.

Indoor Germination Temperature: 65°F

Days to Germination: 21–28

Growing On Temperature: 60°F

Garden Needs: Plant in full sun in rich, fertile, loose, and extremely well-drained soil. Prepare the soil well and add lots of well-aged organic matter. Fertilize the seedlings with a complete, light fertilizer after planting out in the spring. Mulch between the plants and rows. Keep the plants very well watered throughout the growing season, especially while fruits are developing.

Days to Maturity: 65–80

Harvesting Tips: Harvest the fruits when they are fully red. Regular picking will help promote additional fruiting.

Recommended Cultivar: 'Picnic' (everbearing) has a bushy, compact

> **NOTE:** *There are two types of strawberries: those that produce fruit once a year (single crop) and those that produce fruit in the spring and early summer and then again in the late summer and fall (everbearing).*

habit with short runners and produces deliciously sweet medium-size berries. Plants grown from seed will produce fruit the first year in just 4 months, if the seeds are started in late winter.

SWISS CHARD

Beta vulgaris

Sowing Directions: Sow seeds directly outdoors 2 weeks before the last frost. Cover them with ½ inch of soil. Space seeds 6–8 inches apart within the rows, and space the rows 2 feet apart. Thin seedlings within rows to 10–12 inches apart.

Days to Germination: 7–12

Garden Needs: Plant in full sun or light shade in fertile, light, well-drained soil. If the soil is poor, feed the plants every 3–4 weeks with a complete liquid fertilizer high in nitrogen. Keep the plants well watered throughout the growing season.

Days to Maturity: 45–55

Harvesting Tips: Larger leaves can be cut from the base of the main stalk. Younger leaves can remain to mature. If left to grow a second year, these biennials will produce a flower stalk. Although this stalk is edible, cut it off before it has a chance to develop and flower so you'll get more foliage production.

TOMATO

Lycopersicon lycopersicum

Sowing Directions: Sow seeds indoors 6–8 weeks before the last frost. Cover the seeds with ¼ inch of

medium. When the seedlings develop two to four true leaves, transplant them into larger growing on containers. Space determinate seedlings about 3 feet apart within the rows, and space the rows 3–4 feet apart. Space indeterminate seedlings 2½ feet apart within the rows, and space the rows 3 feet apart. Be sure to harden-off seedlings before transplanting them to the garden.

Indoor Germination Temperature: 70–75°F

Days to Germination: 5–7

Growing On Temperature: 60–65°F

Garden Needs: Plant in full sun in fertile, well-drained, deeply prepared soil after the last frost when the soil has warmed up a bit. If you mulch the row areas with black plastic before planting, the soil temperature will increase. Plant the seedlings at least 2 inches deeper in the soil than they were planted in their containers to help anchor the seedlings and stimulate root development. Fertilize the seedlings with a side dressing at planting out time and again every 4 weeks with a light, balanced fertilizer.

Days to Maturity: 49–95, after transplanting to the garden

Harvesting Tips: The beauty of growing tomatoes in the home garden is they can be picked when fully red and ripened, unlike the store-bought fruits, which are picked when green and left to ripen during shipping. Pick tomatoes as they mature to encourage the production of more fruits.

NOTE: *There are two types of tomatoes: determinate and indeterminate. Determinate types, whose fruit ripens all at once, have a bushy habit and need a fair amount of space in the garden. Indeterminate types, whose fruit ripens throughout the season, have a vining habit. Indeterminate types can grow along the ground, but they'll take up less garden space if you provide them with a structure on which to grow.*

Recommended Cultivars:

‘Big Girl’ produces large, smooth, crack-resistant red fruits that can weigh 1 pound or more. It has very good disease-resistance. Matures: 78 days, after transplanting to the garden.

‘Burpee’s Early Pick’ produces bright red, solid fruits all summer. It is resistant to Verticillium and Fusarium wilts. Matures: 62 days, after transplanting to the garden.

The Automator for tomatoes

'Gardener's Delight' produces a very sweet cherry tomato. Matures: 65 days, after transplanting to the garden.

'Heatwave II' is an excellent variety for areas with intense summer 90°F plus heat. It is a disease-resistant variety. Matures: 68 days, after transplanting to the garden.

'Northern Exposure' is a recommended cultivar for areas with short growing seasons. It has excellent disease-resistance. Matures: 67 days, after transplanting to the garden.

TIP:
Turnips are more heat-tolerant than rutabagas.

TURNIP

Brassica rapa

Sowing Directions: This is a cool-weather crop. Sow seeds directly into the garden in early spring as soon as the soil can be worked. Cover them by ½ inch with soil. Space seeds 2 inches apart within the rows, and space the rows 2 feet apart. Thin the seedlings to 4 inches apart with in the rows. For a fall crop, sow seeds directly into the garden in midsummer.

Days to Germination: 7–10

Garden Needs: Plant in full sun in rich, well-dug, moist, well-drained soil with lots of organic matter and compost worked in. Mulch plants to keep them cool and moist. Be sure to keep them consistently well watered.

Days to Maturity: 35–55

Harvesting Tips: Turnips are fast growers. Both their leaves and roots are edible. The leaves can be harvested when young and eaten steamed or simmered for a delicious and nutritious greens dish. The roots can be harvested as soon as they are 2–3 inches; they are eaten raw or cooked.

Recommended Cultivar: 'Tokyo Cross' hybrid is the best hybrid for producing an early spring crop of smooth, white roots. Begin harvesting roots when they are 2 inches across and continue until they are 6 inches across (they'll still taste good). Matures: 35 days.

WATERMELON

Citrullus lanatus

Sowing Directions: This crop needs a long, warm growing season. For best results, sow seeds indoors 4–5 weeks before the last frost in individual peat pots that can be directly transplanted into the garden, to avoid transplant shock. Cover the seeds with 1 inch of soil. In southern climates, seeds can be directly sown into the garden in spring when all danger of frost has passed. Sow five to six seeds per mound at a depth of 1 inch. Space soil mounds (4–6 inches high and 12 inches around) 6 feet apart within the rows, and space the rows 5–7 feet apart. Thin or plant seedlings so that there are two or three per mound.

Indoor Germination Temperature: 70°F

Days to Germination: 6–8

Growing On Temperature: 65–70°F

Garden Needs: After all danger of frost has passed, plant in full sun

in light, very well-drained, sandy soil. If the soil is heavy, add sharp sand to improve drainage. Keep the soil evenly moist, especially when the fruits are developing. Fertilize the plants starting when they reach about 6 inches tall and fertilize again with a side dressing when the plants bloom. Use a 5-10-5 or a 5-10-10 fertilizer.

Days to Maturity: 70–85

Harvesting Tips: Watermelons are ready to be harvested when they twist easily off the stem and have a yellowish color where the melon rind has been resting on the ground. Some gardeners like to tap watermelons to test for ripeness: If it's ripe, it should have a hollow sound and not a soft or dull thudding sound.

Recommended Cultivars:

'Bush Sugar Baby' is a compactly growing plant, which is good for smaller gardens, and pro-

'Bush Sugar Baby' watermelon

duces round, red-fleshed fruits that weigh an average of 12 pounds each. Matures: 80 days.

'Fordhook' hybrid is a vigorous grower that produces juicy, delicious red-fleshed fruits that weigh up to 14 pounds. Matures: 74 days.

6

STARTING HERBS
FROM SEED

TIP

For a beautiful garden composition, plant several different types of basil and let them grow side by side.

Most herbs are very easy to start from seed and grow successfully without too much trouble. The one growing requirement which will ensure their success is extremely well-drained soil. Many herbs are native to Mediterranean regions, which have sandy, gravel-filled, dryish, very well-drained soils. If your soil is heavy and/or wet, plant your herbs in raised beds or in containers.

BASIL

Annual
Ocimum basilicum

Sowing Directions: Sow seeds indoors 6–8 weeks before the last frost. Be sure to harden-off seedlings before moving them to the garden. It is easier to sow seed directly outdoors after the last frost. Lightly cover seeds with ¼ inch of medium.

Indoor Germination Temperature: 70°F

Days to Germination: 7–10

Growing On Temperature: 50–55°F

Garden Planting: Plant in full sun in light, moist (but not wet) soil with good drainage. If the soil is very poor, apply a light complete fertilizer to the seedlings when they are a few inches tall. Pinch off the flower heads as they develop to encourage more foliage growth. Space 1 foot apart.

Description: Most types have a rounded, bushy habit, and produce bright green, scented foliage. One variety, 'Purple Ruffles', has dark bronzy purplish foliage. The herb comes in a range of flavors, from sweet to cinnamon and lemon. The leaves are used in many Italian dishes and tomato sauces. Height: 1½–2 feet.

Herb garden basils

Recommended Cultivars:

'Green Bouquet' has tiny flavorful leaves and is excellent for use as a decorative edging plant. Height: 1 foot.

'Green Ruffles' has large, quilted leaves and is highly recommended for culinary use. Height: 2 feet.

BORAGE

Annual
Borago offinalis

Sowing Directions: Sow seeds indoors 6–8 weeks before the last frost in individual peat pots that can be transplanted directly into the garden. For best results, sow seeds directly into the garden in early spring. Cover seeds with ¼–½ inch of soil. Established plants freely self-sow.

Indoor Germination Temperature: 70°F

Days to Germination: 5–8

Growing On Temperature: 60–65°F

Garden Planting: Plant in full to very light shade in any well-drained soil. Space 2 feet apart.

Description: This herb has a shrub-like habit. Its dark green leaves have a rough texture and slight aroma. It produces lovely, clear blue to violet star-shaped flowers that bloom in summer. Both the leaves and flowers are used to flavor and add color to summer drinks and salads. Height: 2–3½ feet.

BURNET (SALAD)

Perennial
Poterium sanguisorba
Zones: 4–8

Sowing Directions: For best results, sow seeds outdoors, in early spring, directly into the garden where they are to grow. Cover seeds with ½ inch of soil.

Days to Germination: 9–12

Garden Planting: Plant in full sun in rich, moist, well-drained soil. It prefers cool growing conditions. Pinch back flowers to encourage more foliage growth, but leave a few flower heads if you want plants to self-sow. Space 1 foot apart.

Description: This herb is a very attractive plant for any part of the garden. The leaves have a fresh cucumber taste that livens up summer drinks and salads. Height: 15 inches; spread: 15 inches.

CARAWAY

Biennial
Carum carvi
Zones: 3–8

Sowing Directions: Sow seeds directly into the garden in late summer to early fall. Cover the seeds with ½ inch of soil. Plants grown from seed sown in the spring will not produce seed until the following year. Mature plants usually self-sow.

Days to Germination: 10–14

Garden Planting: Plant in full sun in an open area in rich, moist, well-drained soil. Space 12 inches apart.

Description: This herb has fernlike foliage and charming white flowers that resemble Queen Anne's lace. Seeds are harvested in mid to late summer when they turn a brownish color. The mature seeds have a camphorlike smell and are used for flavoring baked goods and stews. The foliage can be used to add flavor to salads, and the edible roots can be steamed. Height: 2 feet.

CATNIP

Perennial
Nepeta cataria
Zones: 3–9

Sowing Directions: Sow seeds indoors 6–8 weeks before the last frost. Can be sown outdoors, in early fall or early spring, directly where they are to grow. Cover with ½ inch of medium.

Indoor Germination Temperature: 70°F

Days to Germination: 5–10

TIP:
Borage is an extremely fast- and vigorous-growing plant that may become invasive if not kept in regular check. To decrease invasiveness, plant it in poor, dry soil or in a good-size container.

Growing On Temperature: 50–55°F

Garden Planting: Plant in full sun in fertile, sandy, well-drained soil. Shear back the plant after flowering for a tidier appearance and to encourage another blooming period. Space 1½ feet apart.

Description: This herb has a mounding habit and produces fragrant bluish green foliage and masses of small blue flowers. The aromatic foliage is used to flavor teas. In addition, of course, it's grown for the undiluted delight of all cats. And if you have a garden, don't forget that cats make some of the best and most interesting of all garden ornamentations! Height: 3½ feet.

CHAMOMILE

Matricaria recutita: (annual) Chamomile, German chamomile; *Chamaemelum nobile:* (perennial) Roman chamomile
Zones: 3–10

Sowing Directions: Sow seeds indoors 6–8 weeks before the last frost. Can be sown directly outdoors in early spring as soon as the soil can be worked. Lightly cover the seed with medium. Mature plants will readily self-sow, if the flowers are not harvested.

Indoor Germination Temperature: 55–60°F

Days to Germination: 10–14

Growing On Temperature: 55°F

Garden Planting: Plant in full sun in dry, well-drained, sandy soil that has some organic matter worked into it. The perennial species prefers an acidic soil. Space 8 inches apart.

NOTE: *Plant out seedlings as soon as possible, since older chamomile plants do not transplant well.*

Description: This herb is grown for its sweet apple-scented foliage and for its delicate, white daisylike flowers that bloom throughout the summer. The flowers are used fresh or dried for teas. Height: 2 feet (annual); 9 inches (perennial).

CHERVIL

Annual
Anthriscus cerefolium

Sowing Directions: This is a cool-weather herb. Sow seed indoors 6–8 weeks before the last frost in individual peat pots that can be transplanted directly into the garden. Can be sown outdoors directly into the garden in early spring 2–3 weeks before the last frost. In warmer regions, seeds can be sown in late summer or early autumn for a fall crop. Lightly cover seeds with ¼ inch of medium.

Indoor Germination Temperature: 65°F

Days to Germination: 7–10

Growing On Temperature: 55°F

Garden Planting: Plant in full sun or light shade (in warmer regions) in rich, moist, soil that has compost and other organic matter worked into it. Space 10 inches apart.

Description: Small white flowers bloom in late spring. It is grown for its lacy, anise-flavored foliage, which is used in salads, soups, stews, and fish dishes. The herb is short lived, completing its total growth cycle in 6–8 weeks. Height: 2 feet; spread: 1 foot.

CHIVES. SEE *ALLIUM* (IN CHAPTER 4)

CICELY (SWEET)

Perennial
Myrrhis odorata
Zones: 3–9

Sowing Directions: For indoor sowing, place seeds in a moist medium and chill for 1–2 months at 40–45°F. Sow cold-treated seed indoors 8–10 weeks before planting out. It is easier to sow seeds outdoors, in early spring or late summer to early fall, directly into the garden where they are to grow. Lightly cover seeds with ¼ inch of medium.

Indoor Germination Temperature: 55–60°F

Days to Germination: 25–30

Growing On Temperature: 55–60°F

Garden Planting: Plant in light shade in rich, moist, humusy soil. Keep the plants well watered. It prefers cool growing conditions. Space 2 feet apart.

Description: The aromatic, bright green foliage is finely cut and fern-like. Lacy white flower clusters bloom in the spring. All parts of this herb—foliage, flowers, seeds, and stems—have an anise flavor and can be used in salads and in fish and seafood dishes. Height: 3 feet; spread: 2 feet.

CORIANDER, CILANTRO, CHINESE PARSLEY

Annual
Coriandrum sativum

Sowing Directions: Sow seeds indoors 4–6 weeks before the last frost in individual peat pots that can be directly transplanted into the garden. Seeds can be sown directly into the garden after the last frost. For a summer-long harvest, sow seeds successively in 3-week intervals. Seeds need darkness to germinate; lightly cover with ¼ inch of medium.

Indoor Germination Temperature: 60°F

Days to Germination: 7–10

Growing On Temperature: 55°F

Garden Planting: Plant in full sun in an open area in loamy, well-drained soil. If the soil is poor, incorporate well-aged compost into the planting area. Keep the plants well watered. Space 6–8 inches apart.

Description: This herb's foliage can be harvested anytime and is used in salsas and other Mexican dishes as well as in many Asian cuisines. The seeds are harvested in midsummer for use in cooking and baking. Height: 2 feet.

CUMIN

Annual
Cuminum cyminum

Sowing Directions: Sow seeds indoors 6–8 weeks before the last frost. Can also be sown directly into the garden after the last frost.

Indoor Germination Temperature: 70°F

Days to Germination: 10–14

Growing On Temperature: 60°F

Garden Planting: After danger of frost has passed, plant in full sun in light, well-drained soil. Space 1–2 feet apart.

Description: This herb has dark green fernlike foliage and white to light pink flowers that bloom in midsummer. The seeds are harvested to use in cooking and baking. It is an essential ingredient in many curry recipes, chutneys, and Mexican dishes. Height: 1–2 feet.

DILL

Annual
Anethum graveolens

Sowing Directions: Sow seed directly into the garden in early spring. The seeds need light to germinate; lightly press seeds onto soil surface and do not cover with soil. Successive sowings can be made until late spring. Mature plants readily self-sow.

Days to Germination: 21

Garden Planting: Plant in full sun in rich, moist soil. Pinch off growing shoots of young plants to encourage more foliage production. Space 1 foot apart.

Description: This herb has fine, feathery, bluish green foliage and greenish yellow flowers that bloom in midsummer. Leaves can be used at anytime in the plant's development for cooking, especially for fish and sauces. Seeds are harvested for use in pickling. Height: 3 feet; spread: 2 feet.

'Fernleaf' dill

Recommended Cultivar: 'Fernleaf' is a compact variety that can be grown in all areas, especially in warmer regions, since it is slow to bolt with the onset of warmer weather. Height: 18 inches.

FENNEL

Annual
Foeniculum vulgare

Sowing Directions: Sow seeds indoors 4–6 weeks before the last frost in individual peat pots that can be transplanted directly into the garden. Can be sown directly outdoors 1–2 weeks before the last frost. Seeds need darkness to germinate; cover seeds with ¼ inch of medium, making sure they are well covered.

Indoor Germination Temperature: 70°F

Days to Germination: 7–10

Growing On Temperature: 55–60°F

Fennel

Garden Planting: Plant in full sun in well-drained, slightly alkaline soil that has lots of organic matter worked into it. Space 12–18 inches apart.

> **NOTE:** *There are two major kinds of fennel: wild and Florence. The seeds and foliage of both types are used, but it's the Florence fennel that produces the thickened stem and bulbous stem base that is harvested and used as a vegetable. It is eaten raw, steamed, or grilled.*

Description: This herb produces fine, needle- or fernlike foliage and greenish yellow umbel flowers that bloom in summer. The anise-flavored seeds and foliage are used in many sauces, salads, and fish dishes. Height: 4–5 feet (wild) or 2½ feet (Florence).

HOREHOUND

Perennial
Marrubium vulgare
Zone: 3

Sowing Directions: Sow seeds indoors 6–8 weeks before the last frost in individual peat pots that can be transplanted directly into the garden. Can be sown outdoors where they are to grow in early spring as soon as the soil can be worked. Cover seeds with ½ inch of medium.

Indoor Germination Temperature: 70°F

Days to Germination: 12

Growing On Temperature: 55–60°F

Garden Planting: Plant in full sun in any soil with good drainage. It thrives in poor soils and is tolerant of heat and drought conditions. Space 12–15 inches apart.

Description: This herb is grown for its minty-tasting, woolly, gray-green leaves, which are used to make teas and to flavor candies, honey, and cough medicines. It bears small white flowers throughout the summer. Height: 28 inches.

HYSSOP

Perennial
Hyssopus officinalis
Zones: 4–9

Sowing Directions: Sow seeds indoors 6–8 weeks before the last frost date. It's easiest to sow seeds directly into the garden in early spring. Cover seeds with ¼ inch of medium.

TIP:
If growing fennel for the foliage rather than its seeds, pinch off the flowers when they appear; this will cause the plant to direct more energy into foliage growth.

Indoor Germination Temperature: 70°F

Days to Germination: 7–10

Growing On Temperature: 60°F

Garden Planting: Plant in full sun to light shade in light, dry, well-drained soil. It is very heat- and drought-tolerant and will do fine even in poor soil. This herb has a tendency to become invasive if not kept in check. Space 2–3 feet apart.

Description: The foliage is dark green and fragrant, upright spikes of white, purple, or violet flowers bloom in early summer and attract all manner of flying garden creatures, especially bees and butterflies. The foliage, which can be used at anytime, is a common ingredient in potpourris and some medications. It can also be used to flavor meats, soups, stews, and liqueurs. Height: 2 feet; spread: 2 feet.

LAVENDER, ENGLISH LAVENDER

Perennial
Lavandula angustifolia
Zones: 6–9

Sowing Directions: To improve germination rates, chill seeds for 4–6 weeks at 35–40°F before sowing. Sow seeds indoors 8–10 weeks before the last frost. Be sure to harden off seedlings before transplanting them to the garden. In Zones 6–9, seeds can be sown directly into the garden in the fall. Seeds need light to germinate; gently press seeds onto the surface and do not cover with medium. In Zones 5–9, seeds can be sown directly into the garden in the fall.

Indoor Germination Temperature: 65–75°F

Days to Germination: 14–21

Growing On Temperature: 60°F

Garden Planting: After the last frost, plant in full sun in dry, well-drained, alkaline soil. After flowering, cut the plant back to keep it tidy and to promote a second flush of flowers. Space 1½ feet apart.

Description: One of the most beautiful of all garden plants with a prized fragrance, this herb has bluish green foliage and gorgeous violet-blue spiky flowers. Both foliage and flowers are used for their soothing, blissful fragrance and and for flavoring teas and some foods (like ice cream, meats, and baked goods). Height: 30 inches.

'Lady' lavender

Recommended Cultivars:

'Hidcote' produces silvery foliage and dark purple flowers and has a compact habit and. Plants grown from seed will not be as uniform as plants propagated by cuttings. Height: 24 inches.

'Lady' is an award-winning, early blooming variety that will flower the first year from seed. It has a compact habit and is excellent for edgings and container plantings. North of Zone 5, grow as an annual. Height: 16 inches.

LEMON BALM

Perennial
Melissa officinalis
Zones: 4–10

Sowing Directions: Sow seeds indoors 4–6 weeks before the last frost. Can be sown outdoors, 3 weeks before the last frost, directly into the garden where they are to grow. Mature plants will freely self-sow. Seeds need light to germinate.

Indoor Germination Temperature: 65°F

Days to Germination: 10–14

Growing On Temperature: 55–60°F

Garden Planting: Plant in full sun to light shade in any soil with good drainage. It tolerates poor soils quite well. Space 2 feet apart.

Description: This fast-growing herb has a spreading habit and bright green foliage with a warm, lemony smell and flavor. Leaves can be used anytime for teas and flavoring drinks. Height: 2 feet.

LOVAGE

Perennial
Levisticum officinale
Zones: 3–9

Sowing Directions: For best results, sow seeds directly into the garden in early fall or early spring. Cover seeds with ¼ inch of soil.

Days to Germination: 10–14

Garden Planting: Plant in full sun to light shade in an open area in rich, well-drained, moist soil. It thrives in soil that has lots of humus and other organic matter mixed into it. Space 2 feet apart.

Description: This erect-growing, shrubby plant has dark green leaves and small yellow flowers that bloom in midsummer. It is grown for its celerylike foliage, seeds, and stalks. Leaves can be cut at anytime to flavor soups and sauces. Height: 3–4 feet.

MARJORAM, SWEET MARJORAM

Annual
Origanum majorana

Sowing Directions: Sow seeds indoors 6–8 weeks before the last frost. Can be sown outdoors, 1–2 weeks before the last frost, directly where they are to grow in the garden. Lightly cover seeds with ¼ inch of medium. Seedlings are susceptible to damping off.

Indoor Germination Temperature: 70°F

TIP:
Lemon balm will become invasive unless regularly kept in check. Container growing is a good option.

TIP:
Mints are notorious for becoming invasive, so keep them in check by growing them in containers or by regularly pulling up unwanted plants.

Days to Germination: 8–14

Growing On Temperature: 55–60°F

Garden Planting: Plant in full sun in light, well-drained, sandy soil with humus or other organic matter worked into it. Space 1–2 feet apart.

Description: A wonderfully fragrant herb that produces small, grayish green leaves that can be harvested anytime. Small white or blue flowers bloom in midsummer. The leaves are used fresh or dried to flavor meats, salads, dressings, soups, and stews. Height: 1–2 feet; spread: 2 feet.

MINT

Perennial
Mentha aquatica var. *crispa*: curly mint, *M.* × *piperita*: peppermint, *M. spicata*: spearmint
Zones: 3–10

Sowing Directions: Sow seeds indoors 4–6 weeks before the last frost. Can be sown directly into the garden in mid to late spring. Lightly cover seeds with ¼ inch of medium.

Indoor Germination Temperature: 70–75°F

Days to Germination: 10–12

Growing On Temperature: 55–60°F

Garden Planting: Plant in full sun to light shade in light, well-drained, moist soil. Provide water during dry periods. Space 2 feet apart.

Description: These are fast-growing herbs with spreading habits and square stems. The leaves can be used anytime to flavor teas and candies. Height: 2–3 feet (curly mint); 2½ feet (peppermint); or 1–2 feet (spearmint).

OREGANO

Perennial
Origanum vulgare
Zones: 3–9

Sowing Directions: Sow seeds indoors 6–8 weeks before the last frost. Can also be directly sown outdoors after the last frost when the soil has warmed up a bit. Lightly cover seeds with ¼ inch of medium.

Indoor Germination Temperature: 70–75°F

Days to Germination: 4–8

Growing On Temperature: 55–60°F

Garden Planting: Plant in full sun in a dryish, slightly alkaline soil after the last frost. Thin or space plants to 10–12 inches apart within rows. Space rows 1–1½ feet apart. Height: 1–2 feet.

Description: Cut leaves anytime before the plants flower. Pinch back flower buds to extend harvest time. Allowing the leaves to dry before using strengthens their spicy flavor. Leaves are used most often in Italian dishes.

PARSLEY

Biennial; grow as an annual
Petroselinum crispum

Sowing Directions: For indoor sowing, sow seeds indoors 6–8 weeks before the last frost in individual peat pots that can be directly transplanted into the garden. Sow several seeds in each pot and thin to the healthiest seedlings. For outdoor sowing, in midspring sow seeds very thinly into the garden where they are to grow. Seeds need darkness to germinate; cover seeds with ¼ inch of very fine medium.

Indoor Germination Temperature: 70°F

Days to Germination: 15–20

Growing On Temperature: 60°F

Garden Planting: Plant in full sun to light shade in light, well-drained soil. For increased foliage production, fertilize at the seedling stage and again every 3–4 weeks with a light, complete fertilizer throughout the growing season. If the plant is left to grow a second season, harvest the leaves before the flowers form; after it flowers, the leaves become bitter. Space 8 inches apart; space rows 2 feet apart.

> **NOTE:** *Parsley comes in three main types:* P. crispum *(curly) and* P. crispum *var.* neapolitanum *(flat-leaved or Italian) are grown for their foliage.* P. crispum *var.* tuberosum *(parsnip-rooted) is grown for its roots.*

Description: This herb has a rounded habit and its leaves and sprigs can be harvested at anytime. Parsnip-rooted parsley is dug in early fall after a few light frosts. Not just a ubiquitous garnish, this herb adds its unique flavor to potatoes, salads, soups, stews, and egg dishes. Height: 10–18 inches.

Recommended Cultivar: 'Extra Curled Dwarf' is an attractive, compact variety with finely cut leaves. Height: 10–12 inches.

PENNYROYAL

Perennial
Mentha pulegium
Zones: 6–10

Sowing Directions: Sow seeds indoors 6–8 weeks before the last frost. Can be sown directly into the garden in early spring when the soil can be worked. Cover seeds with ¼ inch of medium.

Indoor Germination Temperature: 60–70°F

Days to Germination: 12–16

Growing On Temperature: 55–60°F

Garden Planting: Plant in full sun or part shade in well-drained, moist, humusy soil. Space 1 foot apart.

> **NOTE:** *Pennyroyal is not considered safe for human consumption.*

Growth Description: This low-growing, spreading herb has tiny pinkish purple flowers that bloom in late summer and early fall. The leaves have a distinct musky scent (which some people find unappealing) and are used as an insect repellent. Height: 6 inches.

ROSEMARY

Perennial
Rosmarinus officinalis
Zones: 7–10

Sowing Directions: In northern areas, sow seeds indoors 8–10 weeks

TIP: Higher germination rates can be achieved by soaking seeds in warm water for 12–24 hours before sowing. Or try chilling the seeds in a refrigerator for 1–2 days before sowing.

TIP: For Zones north of 7, grow rosemary as an annual or dig up your plants and bring them inside for the winter.

before the last frost. Transplant hardened off seedlings into the garden in spring. In southern areas, seeds can be sown directly outdoors in early spring as soon as soil can be worked. Press the seed into contact with the soil and do not cover.

Sow double or triple the amount of seed you normally would for other plants. To aid germination rates, make sure the seed is fresh and alternate germination temperatures from 70°F during the day to 55°F through the night. Rosemary seeds need perfectly drained soil; do not allow it to get too wet. Just lightly mist the surface of the soil to keep it moist.

Note: *Rosemary can be difficult to start from seed and has low germination rates.*

Indoor Germination Temperature: 70°F (day), 55°F (night)

TIP:

Experienced propagators tell me the secret to starting rosemary is to incorporate a bit of sharp sand or very fine gravel into a 1:1 vermiculite to perlite mixture.

Rosemary

Days to Germination: 18–21

Growing On Temperature: 55°F

Garden Planting: Plant in full sun in loose, sandy, alkaline, gritty, well-drained soil. Space 2–3 feet apart.

Description: This shrubby herb is an evergreen. The aroma of its foliage is one of nature's finest gifts to us. The leaves are used in cooking, especially meat dishes, and for aromatic products such as soaps and oils. Height: 2–5 feet.

RUE

Perennial
Ruta graveolens
Zones: 4–8

Sowing Directions: Sow seeds indoors 6–8 weeks before the last frost. Can be sown directly into the garden in early spring as soon as soil can be worked. Sow seeds ½ inch deep.

Indoor Germination Temperature: 70°F

Days to Germination: 9–12

Growing On Temperature: 55–60°F

Garden Planting: Plant in full sun in an open area in well-drained neutral to slightly alkaline soil. Keep well watered during dry periods; it doesn't thrive in prolonged hot and humid conditions. Pinch back the growing tips of the seedlings when they are 1 foot tall for a fuller habit. Space 1–2 feet apart.

Description: This herb has a wonderful open, airy, rounded habit with blue-gray-green delicately cut foliage. Yellow flowers bloom in midsummer. Height: 2 feet; spread: 1 foot.

NOTE: *Rue was once used as a medicinal herb and as an insect repellent. Now it is used as an ornamental plant. Be careful touching rue. It causes dermatitis in sensitive people.*

Recommended Cultivar: 'Jackman's Blue' is noted for its pronounced bluish-tinged foliage.

SAGE (COMMON)

Perennial
Salvia officinalis
Zones: 4–9

Sowing Directions: Sow seeds indoors 6–8 weeks before the last frost. Can be sown directly into the garden in early spring. Cover the seeds with ½ inch of medium.

Sage

Indoor Germination Temperature: 70°F

Days to Germination: 14–21

Growing On Temperature: 55–60°F

Garden Planting: Plant in full sun in loose, alkaline, well-drained, sandy soil with some well-aged compost or other organic matter worked in. Space 15 inches apart.

Description: This herb has a shrubby habit and commonly has bluish green, soft woolly leaves and small purple flowers that bloom in midsummer. Some cultivars have variegated foliage, and there is also a purple variety. The leaves can be used fresh or dried for flavoring dressings, meats, and vegetable dishes. Height: 2 feet; spread: 2–3 feet.

Recommended Cultivars:

'Aurea' has attractive golden green variegated foliage.

'Ictarina' is a lovely plant with golden variegated leaves. Height: 2–3 feet.

SAVORY

Satureia. hortensis: (annual) summer savory; *S. montana:* (perennial) winter savory.
Zones: 5–10

Sowing Directions: Sow the annual species indoors 6–8 weeks before the last frost and the perennial 4–6 weeks. The perennial can be sown directly into the garden in early to midspring. Seeds need light to germinate; gently press seeds onto the surface and cover with just a dusting of medium.

Indoor Germination Temperature: 70°F

Days to Germination: 12–15 (annual); 15–18 (perennial)

Growing On Temperature: 60°F

Garden Planting: Plant in full sun in an open area in loose, well-drained soil. Pinch back the herb when young to encourage bushiness. Space 18 inches apart.

Description: Both species produce white to pink and purple flowers throughout the summer. Peppery-tasting leaves can be harvested at anytime for use in flavoring salads, meats, vegetables, and bean dishes.

S. hortensis has a loose, open habit. Height: 2 feet.

S. montana has a rounded, compact habit and is usually evergreen. Height: 1 foot.

SORREL (GARDEN, FRENCH)

Perennial
Rumex acetosa
Zones: 4–8

Sowing Directions: Sow seeds indoors 4 weeks before the last frost. Can be directly sown in the garden in early spring as soon as soil can be worked. Sow seeds 1 inch deep in well-worked, moist medium.

Indoor Germination Temperature: 65–70°F

Days to Germination: 12–15

Growing On Temperature: 60°F

Garden Planting: Plant in full sun to light shade in rich, moist soil. Pinch off flowers when they develop to encourage more foliage growth. Keep on eye on this herb, as it can self-sow to invasiveness. Space 2 feet apart.

French sorrel

Description: This herb has an upright, clumping habit. The spicy and tangy foliage has a lemon flavor and is delicious in salads, soups, egg and fish sauces. Height: 30 inches; spread: 24 inches.

TANSY

Perennial
Tanacetum vulgare
Zones: 3–10

Sowing Directions: Sow seeds outdoors, in early spring as soon as the soil can be worked, directly into the garden where they are to grow. Established plants left to flower will readily self-sow. Cover seeds with ¼ inch of soil.

Days to Germination: 10–15

Garden Planting: Plant in full sun in rich, humusy, well-drained soil. Space 2 feet apart.

TIP:
Tansy can become invasive, so keep it in regular check by pulling up unwanted seedlings or removing the flowers before they go to seed.

NOTE: *Tansy is not considered safe for human consumption.*

Description: This ornamental herb is lovely with its dark green, aromatic, fernlike foliage and small, round yellow flowers that bloom in mid to late summer. The flowers are used for dried arrangements. The leaves are used as an insect repellent. Height: 2–3 feet.

TARRAGON

Most often grown as an annual.
Perennial in Zones 8–11
Artemisia dracunculus

Indoor Germination Temperature: 60–70°F

Day to Germination: 15–20

Growing on Temperature: 60°F

Garden Planting: Plant in full sun in any well-drained soil. Tarragon likes a fairly rich soil so add lots of compost or well-aged manure to planting site area. Space out plants 1–1½ feet apart in the garden. Mulch plants over the winter for added protection.

NOTE: *French tarragon (A. dracunculus var. sativa) is usually sold only as plants and propagated by stem or root cuttings. Russian tarragon is available as seed but is considered not as fine for cooking as French tarragon.*

Description: French tarragon grows up to about 2 feet tall and Russian tarragon grows up to 5 feet or more in height. Both have narrow lance-shaped leaves. Tarragon is grown for its use in cooking and flavoring vinegars.

THYME (COMMON, CULINARY)

Perennial
Thymus vulgaris
Zones: 5–9

Sowing Directions: Sow seeds indoors 8 weeks before the last frost. Can also be sown outdoors, 2–3 weeks before the last frost, directly where they are to grow.

Indoor Germination Temperature: 70°F

Days to Germination: 21–30

Growing On Temperature: 55–60°F

Garden Planting: Plant in full sun in an open area in gritty, sandy, well-drained, alkaline soil. It thrives on hot, dry conditions. Water only during an extended drought period. Cut it back in midsummer and early fall for a fuller, tidier habit. Save these cuttings to dry for culinary use. Space 1 foot apart.

Description: This herb's small, blue-green leaves are used fresh or dried in many poultry, pork, vegetable, and salad dishes. Tiny light violet flowers bloom in early summer. Height: 10–12 inches; spread: 1 foot.

WORMWOOD ARTEMISIA

Perennial
Artemesia absinthium
Zones: 4–10

Sowing Directions: Sow seeds outdoors directly in the garden in mid to late spring or fall. Cover seeds with ¼ inch of soil.

Days to Germination: 7–10

Garden Planting: Plant in full sun in any soil with good drainage. Water only during prolonged drought periods; it is quite heat- and drought-tolerant. Space 1½ feet apart.

Description: Grown for its beautiful, silvery green, finely cut, aromatic foliage, which acts as an insect repellent. Height: 4 feet.

NOTE: *Wormwood is considered unsafe for human consumption.*

APPENDIX

AVERAGE DATES OF FIRST AND LAST FROSTS IN THE UNITED STATES

STATE AND CITY	MEAN DATE LAST 32°F IN SPRING	MEAN DATE LAST 32°F IN FALL	MEAN FREEZE FREE DAYS	STATE AND CITY	MEAN DATE LAST 32°F IN SPRING	MEAN DATE LAST 32°F IN FALL	MEAN FREEZE FREE DAYS
ALABAMA				Orlando	Jan. 31	Dec. 17	319
Birmingham	Mar. 19	Nov. 14	241	Pensacola U	Feb. 18	Dec. 15	300
Mobile U	Feb. 17	Dec. 12	298	Tallahassee	Feb. 26	Dec. 3	280
Montgomery U	Feb. 27	Dec. 3	279	Tampa	Jan. 10	Dec. 26	349
ALASKA				**GEORGIA**			
Anchorage	May 18	Sept. 13	118	Atlanta U	Mar. 20	Nov. 19	244
Barrow	June 27	July 5	8	Augusta	Mar. 7	Nov. 22	260
Cordova	May 10	Oct. 2	145	Macon	Mar. 12	Nov. 19	252
Fairbanks	May 24	Aug. 29	97	Savannah	Feb. 21	Dec. 9	291
Juneau U	Apr. 27	Oct. 19	176	**IDAHO**			
Nome	June 12	Aug. 24	73	Boise	Apr. 29	Oct. 16	171
ARIZONA				Pocatello	May 8	Sept. 30	145
Flagstaff	June 8	Oct. 2	116	Salmon	June 4	Sept. 6	94
Phoenix	Jan. 27	Dec. 11	317	**ILLINOIS**			
Tuscon	Mar. 6	Nov. 23	261	Cairo U	Mar. 23	Nov. 11	233
Winslow	Apr. 28	Oct. 21	176	Chicago U	Apr. 19	Oct. 28	192
Yuma U	Jan. 11	Dec. 27	350	Freeport	May 8	Oct. 4	149
ARKANSAS				Peoria	Apr. 22	Oct. 16	177
Fort Smith	Mar. 23	Nov. 9	231	Springfield U	Apr. 8	Oct. 30	206
Little Rock	Mar. 16	Nov. 15	244	**INDIANA**			
CALIFORNIA				Evansville	Apr. 2	Nov. 4	216
Bakersfield	Feb. 14	Nov. 28	287	Fort Wayne	Apr. 24	Oct. 20	179
Eureka U	Jan. 24	Dec. 25	335	Indianapolis U	Apr. 17	Oct. 27	193
Fresno	Feb. 3	Dec. 3	303	South Bend	May 3	Oct. 16	165
Los Angeles U	*	*	*	**IOWA**			
Red Bluff	Feb. 25	Nov. 29	277	Des Moines U	Apr. 20	Oct. 19	183
Sacramento	Jan. 24	Dec. 11	321	Dubuque U	Apr. 19	Oct. 19	184
San Diego	*	*	*	Koekuk	Apr. 12	Oct. 26	197
San Francisco U	*	*	*	Sioux City	Apr. 28	Oct. 12	167
COLORADO				**KANSAS**			
Denver U	May 2	Oct. 14	165	Concordia U	Apr. 16	Oct. 24	191
Palisades	Apr. 22	Oct. 17	178	Dodge City	Apr. 22	Oct. 24	184
Pueblo	Apr. 28	Oct. 12	167	Goodland	May 5	Oct. 9	157
CONNECTICUT				Topeka U	Apr. 9	Oct. 26	200
Hartford	Apr. 22	Oct. 19	180	Wichita	Apr. 5	Nov. 1	210
New Haven	Apr. 15	Oct. 27	195	**KENTUCKY**			
D.C.				Lexington	Apr. 13	Oct. 28	198
Washington U	Apr. 10	Oct. 28	200	Louisville U	Apr. 1	Nov. 7	220
FLORIDA				**LOUISIANA**			
Apalachicola U	Feb. 2	Dec. 21	322	Lake Charles	Feb. 18	Dec. 6	291
Fort Myers	*	*	*	New Orleans	Feb. 13	Dec. 12	302
Jacksonville U	Feb. 6	Dec. 16	313	Shreveport	Mar. 1	Nov. 27	272
Key West	*	*	*	**MAINE**			
Lakeland	Jan. 10	Dec. 25	349	Greenville	May 27	Sept. 20	116
Miami	*	*	*	Portland	Apr. 29	Oct. 15	169

STATE AND CITY	MEAN DATE LAST 32°F IN SPRING	MEAN DATE LAST 32°F IN FALL	MEAN FREEZE FREE DAYS	STATE AND CITY	MEAN DATE LAST 32°F IN SPRING	MEAN DATE LAST 32°F IN FALL	MEAN FREEZE FREE DAYS
MARYLAND				Omaha	Apr. 14	Oct. 20	189
Annapolis	Mar. 4	Nov. 15	225	Valentine Lakes	May 7	Sept. 30	146
Baltimore U	Mar. 28	Nov. 17	234	**NEVADA**			
Frederick	Mar. 24	Oct. 17	176	Elko	June 6	Sept. 3	89
MASSACHUSETTS				Las Vegas	Mar. 13	Nov. 13	245
Boston	Apr. 16	Oct. 25	192	Reno	May 14	Oct. 2	141
Nantucket	Apr. 12	Nov. 16	219	Winnemucca	May 18	Sept. 21	125
MICHIGAN				**NEW HAMPSHIRE**			
Alpena U	May 6	Oct. 9	156	Concord	May 11	Sept. 30	142
Detroit	Apr. 25	Oct. 23	181	**NEW JERSEY**			
Escanaba U	May 14	Oct. 6	145	Cape May	Apr. 4	Nov. 15	225
Grand Rapids U	Apr. 25	Oct. 27	185	Trenton U	Apr. 8	Nov. 5	211
Marquette U	May 14	Oct. 17	156	**NEW MEXICO**			
S. Ste. Marie	May 18	Oct. 3	138	Albuquerque	Apr. 16	Oct. 29	196
MINNESOTA				Rosewell	Apr. 9	Nov. 2	208
Albert Lee	May 3	Oct. 6	156	**NEW YORK**			
Big Falls R.S.	June 4	Sept. 7	95	Albany	Apr. 27	Oct. 13	169
Brainerd	May 16	Sept. 24	131	Binghamptom U	May 4	Oct. 6	154
Duluth	May 22	Sept. 24	125	Buffalo	Apr. 30	Oct. 25	179
Minneapolis	Apr. 30	Oct. 13	166	New York U	Apr. 7	Nov. 12	219
St. Cloud	May 9	Sept. 29	144	Rochester	Apr. 28	Oct. 21	176
MISSISSIPPI				Syracuse	Apr. 30	Oct. 15	168
Jackson	Mar. 10	Nov. 13	248	**NORTH CAROLINA**			
Meridian	Mar. 13	Nov. 14	246	Asheville UE	Apr. 12	Oct. 24	195
Vicksburg U	Mar. 8	Nov. 15	252	Charlotte U	Mar. 21	Nov. 15	239
MISSOURI				Greenville	Mar. 28	Nov. 5	222
Columbia	Apr. 9	Oct. 24	198	Hatteras	Feb. 25	Dec. 18	296
Kansas City	Apr. 5	Oct. 31	210	Raleigh U	Feb. 25	Nov. 16	237
St. Louis	Apr. 2	Nov. 8	220	Wilmington U	Mar. 8	Nov. 24	262
Springfield	Apr. 10	Oct. 31	203	**NORTH DAKOTA**			
MONTANA				Bismarck	May 11	Sept. 24	136
Billings	May 15	Sept. 24	132	Devils Lake U	May 18	Sept. 22	127
Glasgow U	May 19	Sept. 20	124	Fargo	May 13	Sept. 27	137
Great Falls	May 14	Sept. 26	135	Williston U	May 14	Sept. 23	132
Havre U	May 9	Sept. 23	138	**OHIO**			
Helena	May 12	Sept. 23	134	Akron-Canton	Apr. 29	Oct. 20	173
Kalispell	May 12	Sept. 23	134	Cincinnati (ABBE)	Apr. 15	Oct. 25	192
Miles City	May 5	Oct. 3	150	Cleveland	Apr. 21	Nov. 2	195
Superior	June 5	Aug. 30	85	Columbus U	Apr. 17	Oct. 30	196
NEBRASKA				Dayton	Apr. 20	Oct. 21	184
Grand Island	Apr. 29	Oct. 6	160	Toledo	Apr. 24	Oct. 25	184
Lincoln	Apr. 20	Oct. 17	180	**OKLAHOMA**			
Norfolk	May 4	Oct. 3	152	Okla. City U	Mar. 28	Nov. 7	223
North Platte	Apr. 30	Oct. 7	160	Tulsa	Mar. 31	Nov. 2	216

STATE AND CITY	MEAN DATE LAST 32°F IN SPRING	MEAN DATE LAST 32°F IN FALL	MEAN FREEZE FREE DAYS	STATE AND CITY	MEAN DATE LAST 32°F IN SPRING	MEAN DATE LAST 32°F IN FALL	MEAN FREEZE FREE DAYS
OREGON				Midland	Apr. 3	Nov. 6	218
Astoria	Mar. 18	Nov. 24	251	Mission	Jan. 30	Dec. 21	325
Bend	June 17	Aug. 17	62	Mount Pleasant	Mar. 23	Nov. 12	233
Medford	Apr. 25	Oct. 20	178	Nacodoches	Mar. 15	Nov. 13	243
Pendleton	Apr. 27	Oct. 8	163	Plainview	Apr. 10	Nov. 6	211
Portland U	Feb. 25	Dec. 1	279	Presidio	Mar. 20	Nov. 13	238
Salem	Apr. 14	Oct. 27	197	Quanah	Mar. 31	Nov. 7	221
PENNSYLVANIA				San Angelo	Mar. 25	Nov. 15	235
Allentown	Apr. 20	Oct. 16	180	Ysleta	Apr. 6	Oct. 30	207
Harrisburg	Apr. 10	Oct. 28	201	**UTAH**			
Philadelphia U	Mar. 30	Nov. 17	232	Blanding	May 18	Oct. 14	148
Pittsburgh	Apr. 20	Oct. 23	187	Salt Lake City	Apr. 12	Nov. 1	202
Scranton U	Apr. 24	Oct. 14	174	**VERMONT**			
RHODE ISLAND				Burlington	May 8	Oct. 3	148
Providence U	Apr. 13	Oct. 27	197	**VIRGINIA**			
SOUTH CAROLINA				Lynchburg	Apr. 6	Oct. 27	205
Charleston U	Feb. 19	Dec. 10	294	Norfolk U	Mar. 18	Nov. 27	254
Columbia U	Mar. 14	Nov. 21	252	Richmond U	Apr. 2	Nov. 8	220
Greenville	Mar. 23	Nov. 17	239	Roanoke	Apr. 20	Oct. 24	187
SOUTH DAKOTA				**WASHINGTON**			
Huron U	May 4	Sept. 30	149	Bumping Lake	June 17	Aug. 16	60
Rapid City U	May 7	Oct. 4	150	Seattle U	Feb. 23	Dec. 1	281
Sioux FallsU	May 5	Oct. 3	152	Spokane	Apr. 20	Oct. 12	175
TENNESSE				Tatoosh Island	Jan. 25	Dec. 20	329
Chattanooga U	Mar. 26	Nov. 10	229	Walla Walla U	Mar. 28	Nov. 1	218
Knoxville	Mar. 31	Nov. 6	220	Yakima	Apr. 19	Oct. 15	179
Memphis U	Mar. 20	Nov. 12	237	**WEST VIRGINIA**			
Nashville U	Mar. 28	Nov. 7	224	Charleston	Apr. 18	Oct. 28	193
TEXAS				Parkersburg	Apr. 16	Oct. 21	189
Albany	Mar. 30	Nov. 9	224	**WISCONSIN**			
Balmorhea	Apr. 1	Nov. 12	226	Green Bay	May 6	Oct. 13	161
Beeville	Feb. 21	Dec. 6	288	La Crosse U	May 1	Oct. 8	161
College Station	Mar. 1	Dec. 1	275	Madison U	Apr. 26	Oct. 19	177
Corsicana	Mar. 13	Nov. 27	259	Milwaukee U	Apr. 20	Oct. 25	188
Dallas	Mar. 18	Nov. 22	249	**WYOMING**			
Der Rio	Feb. 12	Dec. 9	300	Casper	May 18	Sept. 25	130
Eencinal	Feb. 15	Dec. 12	301	Cheyenne	May 20	Sept. 27	130
Houston	Feb. 5	Dec. 11	309	Lander	May 15	Sept. 20	128
Lampasas	Apr. 1	Nov. 10	223	Sheridan	May 21	Sept. 21	123
Matagorda	Feb. 12	Dec. 17	308				

Source: United States Department of Agriculture

INDOOR AND OUTDOOR PLANTING GUIDE

Complete instructions are on each seed packet.	SOW	LIGHT	BLOOM
FLOWERS			
Achillea (P)	✳	○	summer–fall
Acroclinium (A)	☼	○	summer–fall
Ageratum (A)	[indoors]	○ ◐	summer–fall
Alyssum (A)	✳	○ ◐	summer–fall
Amaranthus (A)	[indoors]	○	summer
Aster (A)	[indoors] or ✳	○	summer–fall
Baby's breath (A, P)	☼	○	spring–summer
Bachelor's buttons (A)	[indoors] or ✳	○	summer
Balsam (A)	☼	○ ◐	summer–fall
Cabbage, flowering (A)	✳	○	fall
Calendula (A)	✳	○	spring–fall
Candytuft (A)	✳	○	spring–summer
Canterbury bells (B)	☼	○ ◐	spring–summer
Carnation (A)	[indoors]	○	summer–fall
Celosia (A)	[indoors] or ✳	○	summer–fall
Chinese lanterns (P)	[indoors] or ✳	○	fall
Cleome (A)	[indoors] or ✳	○ ◐	summer–fall
Columbine (P)	[indoors] or ✳	◐ ○	spring–summer
Coreopsis (P)	☼	○	spring–fall
Cosmos (A)	☼	○	summer–fall
Daisy, African (A)	[indoors] or ✳	○	summer–fall
Daisy, gloriosa (P)	[indoors] or ✳	○ ◐	summer–fall
Daisy, painted (P)	[indoors] or ☼	○ ◐	summer
Daisy, shasta (P)	[indoors] or ☼	○	summer
Delphinium (P)	[indoors]	○	spring–summer
Dianthus (A)	[indoors]	○	summer–fall
Dusty miller (A)	[indoors] or ☼	○	summer–fall
Forget-me-not (A)	☼	○ ◐	spring
Foxglove (B)	☼	◐ ○	spring–summer
Geranium (A)	[indoors]	○	summer–fall
Globe amaranth (A)	[indoors]	○	summer–fall
Gourd (A)	☼	○	fall
Hollyhock (P)	[indoors] or ☼	○	summer

Complete instructions are on each seed packet.	SOW	LIGHT	BLOOM
Honesty (**B**)	☼	○ ◑	spring–fall
Impatiens (**A**)	⊔⊔	◑ ○	summer–fall
Kale, flowering (**A**)	✳	○	fall
Larkspur (**A**)	✳	○	summer–fall
Lobelia (**A**)	⊔⊔ or ☼	○ ◑	spring–fall
Lupine (**P**)	⊔⊔ or ☼	○ ◑	spring–summer
Marigold (**A**)	⊔⊔ or ☼	○	summer–fall
Mixed annuals (**A**)	⊔⊔ or ☼	○ ◑ ●	spring–fall
Moonflower (**A**)	⊔⊔ or ☼	○	summer–fall
Morning glory (**A**)	⊔⊔ or ☼	○	summer–fall
Nasturtium (**A**)	☼	○	summer–fall
Nicotiana (**A**)	⊔⊔ or ☼	○ ◑	summer–fall
Pansy (**A**)	⊔⊔	◑ ○	spring–summer
Petunia (**A**)	⊔⊔	○	summer–fall
Phlox (**A**)	⊔⊔ or ☼	○	summer–fall
Poppy, California (**A**)	✳	○	spring–summer
Poppy, Oriental (**P**)	⊔⊔ or ☼	○	spring–summer
Portulaca (**A**)	⊔⊔ or ☼	○	summer–fall
Salvia (**A, P**)	⊔⊔ or ☼	○ ◑	summer–fall
Snapdragon (**A**)	⊔⊔ or ☼	○ ◑	summer–fall
Starflower (**A**)	⊔⊔ or ☼	○	summer–fall
Statice (**A**)	⊔⊔	○	summer
Stock (**A**)	⊔⊔ or ✳	○	spring–summer
Strawflower (**A**)	⊔⊔ or ✳	○	summer–fall
Sweet pea (**A**)	⊔⊔ or ✳	○	spring
Sweet William (**B**)	☼	○	spring–summer
Texas bluebonnet (**A**)	☼	○	summer
Vinca (**A**)	⊔⊔	○ ◑	summer–fall
Viola (**P**)	⊔⊔ or ☼	◑ ○	spring–summer
Wildflowers (**A, B, P**)	☼	○	spring–fall
Xeranthemum (**A**)	☼	○	summer
Zinnia (**A**)	⊔⊔ or ☼	○	summer–fall

Key:
⊔⊔ = Start early indoors   ○ = Full Sun
✳ = Start outdoors after danger of heavy frost   ◑ = Partial Shade
✿ = Start outdoors after all danger of frost   ● = Shade
A = Annual
B = Biennial
P = Perennial

Complete instructions are on each seed packet.	SOW	LIGHT	HARVEST
VEGETABLES			
Bean, bush lima	✿	○	70–75 days
Bean, bush snap	✿	○	50–60 days
Bean, pole green	✿	○	50–70 days
Bean, pole lima	✿	○	85–92 days
Beet	✳	○	59–80 days
Broccoli	⊔⊔ or ✳	○	55–80 days
Brussels sprouts	⊔⊔ or ✳	○	90 days
Cabbage	⊔⊔ or ✳	○ ◑	60–76 days
Cabbage, Chinese	⊔⊔ or ✿	○ ◑	43–62 days
Cantaloupe	✿	○	75–90 days
Carrot	✳	○	60–75 days
Cauliflower	⊔⊔ or ✳	○	52–58 days
Celery	⊔⊔	○	105 days
Chicory (radicchio)	✿	○	85 days
Collard	✳	○	60 days
Corn, popcorn	✿	○	90 days
Corn, sweet	✿	○	63–92 days
Cowpea	✿	○	75–78 days
Cucumber	✿	○	52–70 days
Eggplant	⊔⊔	○	55–70 days
Endive	✳	○	90 days
Kale	✳	○	55 days
Kohlrabi	✳	○	55 days
Leek	✳	○	130 days
Lettuce, head	✳	○ ◑	70–85 days
Lettuce, leaf	✳	○ ◑	45–50 days
Mustard	✳	○	45 days
Okra	✿	○	48–56 days

Complete instructions are on each seed packet.	SOW	LIGHT	HARVEST
Onion	(tray) or ✳	○	60–170 days
Pak choi	✳	○ ◑	47 days
Parsnip	✳	○	105 days
Pea	✳	○	61–68 days
Pea, snap	✳	○	56–70 days
Pea, snow	✳ (tray)	○	58–65 days
Pepper, hot	(tray)	○	70–95 days
Pepper, sweet	(tray) ☼	○	65–75 days
Pumpkin	☼	○	95–120 days
Radish	✳	○ ◑	22–45 days
Roquette (arugula)	✳ ☼	○ ◑	35 days
Rutabaga	☼	○	90 days
Spinach	✳	○ ◑	42–48 days
Squash, summer	☼	○	50–57 days
Squash, winter	☼	○	75–100 days
Sunflower	☼	○	80 days
Swiss chard	✳	○ ◑	60 days
Tomato	(tray)	○	52–80 days
Turnip	☼	○	50–70 days
Watermelon	☼	○	80–85 days

Complete instructions are on each seed packet.	SOW	LIGHT	HEIGHT
HERBS			
Basil, lemon (A)	☼	○	12–18"
Basil, ruffles (A)	☼	○	18"
Basil, sweet (A)	☼	○	12–18"
Catnip (P)	☼	○ ◑	18"
Chives, common (P)	☼	○	12–18"
Chives, garlic (P)	☼	○	18–24"
Coriander (cilantro) (A)	☼	○	18–24"
Dill (A)	☼	○	24–36"
Dill, fernleaf (A)	☼	○	18"
Fennel, Florence (A)	✳	○	30"
Lavender (P)	⊎	○	30"
Lemon mint (A)	☼	○ ◑	24"
Oregano (P)	☼	○	12–24"
Parsley, curled (B)	⊎ or ☼	○	12"
Parsley, Hamburg (B)	☼	○	15–18"
Parsley, plain (B)	⊎ or ☼	○	15–18"
Sage (P)	☼	○	12–24"
Savory, summer (A)	☼	○	12"
Spearmint (P)	⊎	○	18–24"
Thyme (P)	⊎	○	6–12"

Source: W. Atlee Burpee & Company

SPRING PLANTING GUIDE FOR VEGETABLES AND HERBS

Crop	Planting dates for localities in which average date of last freeze is—		
	January 30	February 8	February 18
Asparagus [1]			
Beans, lima	Feb. 1 – Apr. 15	Feb. 10 – May 1	Mar. 1 – May 1
Beans, snap	Feb. 1 – Apr. 1	Feb. 1 – May 1	Mar. 1 – May 1
Beet	Jan. 1 – Mar. 15	Jan. 10 – Mar. 15	Jan. 20 – Apr. 1
Broccoli, sprouting [1]	Jan. 1 – 30	Jan. 1 – 30	Jan. 15 – Feb. 15
Brussels sprouts [1]	Jan. 1 – 30	Jan. 1 – 30	Jan. 15 – Feb. 15
Cabbage [1]	Jan. 1 – 15	Jan. 1 – Feb. 10	Jan. 1 – Feb. 25
Cabbage, Chinese	(2)	(2)	(2)
Carrot	Jan. 1 – Mar. 1	Jan. 1 – Mar. 1	Jan. 15 – Mar. 1
Cauliflower [1]	Jan. 1 – Feb. 1	Jan. 1 – Feb. 1	Jan. 10 – Feb. 10
Celery and celeriac	Jan. 1 – Feb. 1	Jan. 10 – Feb. 10	Jan. 20 – Feb. 20
Chard	Jan. 1 – Apr. 1	Jan. 10 – Apr. 1	Jan. 20 – Apr. 15
Chervil and chives	Jan. 1 – Feb. 1	Jan. 1 – Feb. 1	Jan. 1 – Feb. 1
Chicory, witloof			
Collards [1]	Jan. 1 – Feb. 15	Jan. 1 – Feb. 15	Jan. 1 – Mar. 15
Cornsalad	Jan. 1 – Feb. 15	Jan. 1 – Feb. 15	Jan. 1 – Mar. 15
Corn, sweet	Jan. 1 – Mar. 15	Feb. 10 – Apr. 1	Feb. 20 – Apr. 15
Cress, upland	Jan. 1 – Feb. 1	Jan. 1 – Feb. 15	Jan. 15 – Feb.15
Cucumber	Feb. 15 – Mar. 15	Feb. 15 – Apr. 1	Feb. 15 – Apr. 15
Eggplant [1]	Jan. 1 – Mar. 1	Feb. 10 – Mar. 15	Feb. 20 – Apr. 1
Endive	Jan. 1 – Mar. 1	Jan. 1 – Mar. 1	Jan. 15 – Mar. 1
Fennel, Florence	Jan. 1 – Mar. 1	Jan. 1 – Mar. 1	Jan. 15 – Mar. 1
Garlic	(2)	(2)	(2)
Horseradish [1]			
Kale	Jan. 1 – Feb. 1	Jan. 10 – Feb. 1	Jan. 20 – Feb. 10
Kohlrabi	Jan. 1 – Feb. 1	Jan. 10 – Feb. 1	Jan. 20 – Feb. 10
Leek	Jan. 1 – Feb. 1	Jan. 1 – Feb. 1	Jan. 1 – Feb. 15
Lettuce, head [1]	Jan. 1 – Feb. 1	Jan. 1 – Feb. 1	Jan. 1 – Feb. 1
Lettuce, leaf	Jan. 1 – Feb. 1	Jan. 1 – Feb. 1	Jan. 1 – Mar. 15
Muskmelon	Feb. 15 – Mar. 15	Feb. 15 – Apr. 1	Feb. 15 – Apr. 15
Mustard	Jan. 1 – Mar. 1	Jan. 1 – Mar. 1	Feb. 1 – Apr. 15
Okra	Feb. 15 – Apr. 1	Feb. 15 – Apr. 15	Mar. 1 – June 1
Onion [1]	Jan. 1 – 15	Jan. 1 – 15	Jan. 1 – 15
Onion, seed	Jan. 1 – 15	Jan. 1 – 15	Jan. 1 – 15
Onion, sets	Jan. 1 – 15	Jan. 1 – 15	Jan. 1 – 15
Parsley	Jan. 1 – 30	Jan. 1 – 30	Jan. 1 – 30
Parsnip			Jan. 1 – Feb. 1
Peas, garden	Jan. 1 – Feb. 15	Jan. 1 – Feb. 15	Jan. 1 – Mar. 1
Peas, black-eye	Feb. 15 – May 1	Feb. 15 – May 15	Mar. 1 – June 15
Pepper [1]	Feb. 1 – Apr. 1	Feb. 15 – Apr. 15	Mar. 1 – May 1
Potato	Jan. 1 – Feb. 15	Jan. 1 – Feb. 15	Jan. 15 – Mar. 1
Radish	Jan. 1 – Apr. 1	Jan. 1 – Apr. 1	Jan. 1 – Apr.1
Rhubarb [1]			
Rutabaga			
Salsify	Jan. 1 – Feb. 1	Jan. 10 – Feb. 10	Jan. 15 – Feb. 20
Shallot	Jan. 1 – Feb. 1	Jan. 1 – Feb. 10	Jan. 1 – Feb. 20
Sorrel	Jan. 1 – Mar. 1	Jan. 1 – Mar. 1	Jan. 15 – Mar. 1
Soybean	Mar. 1 – June 30	Mar. 1 – June 30	Mar. 10 – June 30
Spinach	Jan. 1 – Feb. 15	Jan. 1 – Feb. 15	Jan. 1 – Mar. 1
Spinach, New Zealand	Feb. 1 – Apr. 15	Feb. 15 – Apr. 15	Mar. 1 – Apr. 15
Squash, summer	Feb. 1 – Apr. 15	Feb. 15 – Apr. 15	Mar. 1 – Apr. 15
Sweet potato	Feb. 15 – May 15	Mar. 1 – May 15	Mar. 20 – June 1
Tomato	Feb. 1 – Apr. 1	Feb. 20 – Apr. 10	Mar. 1 – Apr. 20
Turnip	Jan. 1 – Mar. 1	Jan. 1 – Mar. 1	Jan. 10 – Mar. 1
Watermelon	Feb. 15 – Mar. 15	Feb. 15 – Apr. 1	Feb. 15 – Apr. 15

[1] Plants

[2] Generally fall planted

Planting dates for localities in which average date of last freeze is—

February 28	March 10	March 20	March 30
	Jan. 1 – Mar. 1	Feb. 1 – Mar. 10	Feb. 15 – Mar. 20.
Mar. 15 – June 1	Mar. 20 – June 1	Apr. 1 – June 15	Apr. 15 – June 20.
Mar. 10 – May 15	Mar. 15 – May 15	Mar. 15 – May 25	Apr. 1 – June 1.
Feb. 1 – Apr. 15	Feb. 15 – June 1	Feb. 15 – May 15	Mar. 1 – June 1.
Feb. 1 – Mar. 1	Feb. 15 – Mar. 15	Feb. 15 – Mar. 15	Mar. 1 – 20.
Feb. 1 – Mar. 1	Feb. 15 – Mar. 15	Feb. 15 – Mar. 15	Mar. 1 – 20.
Jan. 15 – Feb. 25	Jan. 25 – Mar. 1	Feb. 1 – Mar. 1	Feb. 15 – Mar. 10.
(2)	(2)	(2)	(2)
Feb. 1 – Mar. 1	Feb. 10 – Mar. 15	Feb. 15 – Mar. 20	Mar. 1 – Apr. 10.
Jan. 20 – Feb. 20	Feb. 1 – Mar. 1	Feb. 10 – Mar. 10	Feb. 20 – Mar. 20.
Feb. 1 – Mar. 1	Feb. 20 – Mar. 20	Mar. 1 – Apr. 1	Mar. 15 – Apr. 15.
Feb. 1 – May 1	Feb. 15 – May 15	Feb. 20 – May 15	Mar. 1 – May 25.
Jan. 15 – Feb. 15	Feb. 1 – Mar. 1	Feb. 10 – Mar. 10	Feb. 15 – Mar. 15.
	June 1 – July 1	June 1 – July 1	June 1 – July 1.
Jan. 15 – Mar. 15	Feb. 1 – Apr. 1	Feb. 15 – May 1	Mar. 1 – June 1.
Jan. 1 – Mar. 1	Jan. 1 – Mar. 15	Jan. 1 – Mar. 15	Jan. 15 – Mar. 15.
Mar. 1 – Apr. 15	Mar. 10 – Apr. 15	Mar. 15 – May 1	Mar. 25 – May 15.
Feb. 1 – Mar. 1	Feb. 10 – Apr. 15	Feb. 20 – Mar. 15	Mar. 1 – Apr. 1.
Mar. 1 – Apr. 15	Mar. 15 – Apr. 15	Apr. 1 – May 1	Apr. 10 – May 15.
Mar. 10 – Apr. 15	Mar. 15 – Apr. 15	Apr. 1 – May 1	Apr. 15 – May 15.
Feb. 1 – Mar. 1	Feb. 15 – Mar. 15	Mar. 1 – Apr. 1	Mar. 10 – Apr. 10.
Feb. 1 – Mar. 1	Feb. 15 – Mar. 15	Mar. 1 – Apr. 1	Mar. 10 – Apr. 10.
(2)	(2)	Feb. 1 – Mar. 1	Feb. 10 – Mar. 10.
			Mar. 1 – Apr. 1.
Feb. 1 – 20	Feb. 10 – Mar. 1	Feb. 20 – Mar. 10	Mar. 1 – 20.
Feb. 1 – 20	Feb. 10 – Mar. 1	Feb. 20 – Mar. 10	Mar. 1 – Apr. 1.
Jan. 15 – Feb. 15	Jan. 25 – Mar. 1	Feb. 1 – Mar. 1	Feb. 15 – Mar. 15.
Jan. 15 – Feb. 15	Feb. 1 – 20	Feb. 15 – Mar. 10	Mar. 1 – 20.
Jan. 1 – Mar. 15	Jan. 15 – Apr. 1	Feb. 1 – Apr. 1	Feb. 15 – Apr. 15.
Mar. 1 – Apr. 15	Mar. 15 – Apr. 15	Apr. 1 – May 1	Apr. 10 – May 15.
Feb. 1 – Mar. 1	Feb. 10 – Mar. 15	Feb. 20 – Apr. 1	Mar. 1 – Apr. 15.
Mar. 10 – June 1	Mar. 20 – June 1	Apr. 1 – June 15	Apr. 10 – June 15.
Jan. 1 – Feb. 1	Jan. 15 – Feb. 15	Feb. 10 – Mar. 10	Feb. 15 – Mar. 15.
Jan. 1 – Feb. 15	Feb. 1 – Mar. 1	Feb. 10 – Mar. 10	Feb. 20 – Mar. 15.
Jan. 1 – Mar. 1	Jan. 15 – Mar. 10	Feb. 1 – Mar. 20	Feb. 15 – Mar. 20.
Jan. 15 – Mar. 1	Feb. 1 – Mar. 10	Feb. 15 – Mar. 15	Mar. 1 – Apr. 1.
Jan. 15 – Feb. 15	Jan. 15 – Mar. 1	Feb. 15 – Mar. 15	Mar. 1 – Apr. 1.
Jan. 15 – Mar. 1	Jan. 15 – Mar. 15	Feb. 1 – Mar. 15	Feb. 10 – Mar. 20.
Mar. 10 – June 20	Mar. 15 – July 1	Apr. 1 – July 1	Apr. 15 – July 1.
Mar. 15 – May 1	Apr. 1 – June 1	Apr. 10 – June 1	Apr. 15 – June 1.
Jan. 15 – Mar. 1	Feb. 1 – Mar. 1	Feb. 10 – Mar. 15	Feb. 20 – Mar. 20.
Jan. 1 – Apr. 1	Jan. 1 – Apr. 15	Jan. 20 – May 1	Feb. 15 – May 1.
Jan. 1 – Feb. 1	Jan. 15 – Feb. 15	Jan. 15 – Mar. 1	Feb. 1 – Mar. 1.
Jan. 15 – Mar. 1	Feb. 1 – Mar. 1	Feb. 15 – Mar. 1	Mar. 1 – 15.
Jan. 1 – Mar. 1	Jan. 15 – Mar. 1	Feb. 1 – Mar. 10	Feb. 15 – Mar. 15.
Feb. 1 – Mar. 10	Feb. 10 – Mar. 15	Feb. 10 – Mar. 20	Feb. 20 – Apr. 1.
Mar. 20 – June 30	Apr. 10 – June 30	Apr. 10 – June 30	Apr. 20 – June 30.
Jan. 1 – Mar. 1	Jan. 15 – Mar. 10	Jan. 15 – Mar. 15	Feb. 1 – Mar. 20
Mar. 15 – May 15	Mar. 20 – May 15	Apr. 1 – May 15	Apr. 10 – June 1.
Mar. 15 – May 15	Mar. 15 – May 1	Apr. 1 – May 15	Apr. 10 – June 1.
Mar. 20 – June 1	Apr. 1 – June 1	Apr. 10 – June 1	Apr. 10 – June 1.
Mar. 10 – May 1	Apr. 20 – May 10	Apr. 1 – May 20	Apr. 10 – June 1.
Jan. 20 – Mar. 1	Feb. 1 – Mar. 1	Feb. 10 – Mar. 10	Feb. 20 – Mar. 20.
Mar. 1 – Apr. 15	Mar. 15 – Apr. 15	Apr. 1 – May 1	Apr. 10 – May 15.

SPRING PLANTING GUIDE FOR VEGETABLES AND HERBS (CONTINUED)

Crop	Planting dates for localities in which average date of last freeze is—		
	April 10	April 20	April 30
Asparagus [1]	Mar. 10 – Apr. 10	Mar. 15 – Apr. 15	Mar. 20 – Apr. 15
Beans, lima	Apr. 1 – June 30	May 1 – June 20	May 15 – June 15
Beans, snap	Mar. 10 – June 30	Apr. 25 – June 30	May 10 – June 30
Beet	Mar. 10 – June 1	Mar. 20 – June 1	Apr. 1 – June 15
Broccoli, sprouting [1]	Mar. 15 – Apr. 15	Mar. 25 – Apr. 20	Apr. 1 – May 1
Brussel sprouts [1]	Mar. 15 – Apr.15	Mar. 25 – Apr. 20	Apr. 1 – May 1
Cabbage [1]	Mar. 1 – Apr. 1	Mar. 10 – Apr. 1	Mar. 15 – Apr. 10
Cabbage, Chinese	(2)	(2)	(2)
Carrot	Mar. 10 – Apr. 20	Apr. 1 – May 15	Apr. 10 – June 1
Cauliflower [1]	Mar. 1 – Mar. 20	Mar. 15 – Apr. 20	Apr. 10 – May 10
Celery and celeriac	Apr. 1 – Apr. 20	Apr. 1 – May 1	Apr. 15 – May 1
Chard	Mar. 15 – June 15	Apr. 1 – June 15	Apr. 15 – June 15
Chervil and chives	Mar. 1 – Apr. 1	Mar. 10 – Apr. 10	Mar. 20 – Apr. 20
Chicory, witloof	June 10 – July 1	June 15 – July 1	June 15 – July 1
Collards [1]	Mar. 1 – June 1	Mar. 10 – June 1	Apr. 1 – June 1
Cornsalad	Feb. 1 – Apr. 1	Feb. 15 – Apr. 15	Mar. 1 – May 1
Corn, sweet	Apr. 10 – June 1	Apr. 25 – June 1	May 10 – June 15
Cress, upland	Mar. 10 – Apr. 15	Mar. 20 – May 1	Apr. 10 – May 10
Cucumber	Apr. 20 – June 1	May 1 – June 15	May 15 – June 15
Eggplant [1]	May 1 – June 1	May 10 – June 1	May 15 – June 10
Endive	Mar. 15 – Apr. 15	Mar. 25 – Apr. 15	Apr. 1 – May 1
Fennel, Florence	Mar. 15 – Apr. 15	Mar. 25 – Apr. 15	Apr. 1 – May 1
Garlic	Feb. 20 – Mar. 20	Mar. 10 – Apr. 1	Mar. 15 – Apr. 15
Horseradish [1]	Mar. 10 – Apr. 10	Mar. 20 – Apr. 20	Apr. 1 – 30
Kale	Mar. 10 – Apr. 1	Mar. 20 – Apr. 10	Apr. 1 – 20
Kohlrabi	Mar. 10 – Apr. 10	Mar. 20 – May 1	Apr. 1 – May 10
Leek	Mar. 1 – Apr. 1	Mar. 15 – Apr. 15	Apr. 1 – May 1
Lettuce, head [1]	Mar. 10 – Apr. 1	Mar. 20 – Apr. 15	Apr. 1 – May 1
Lettuce, leaf	Mar. 15 – May 15	Mar. 20 – May 15	Apr. 1 – June 1
Muskmelon	Apr. 20 – June 1	May 1 – June 15	May 15 – June 15
Mustard	Mar. 10 – Apr. 20	Mar. 20 – May 1	Apr. 1 – May 10
Okra	Apr. 20 – June 15	May 1 – June 1	May 10 – June 1
Onion [1]	Mar. 1 – Apr. 1	Mar. 15 – Apr. 10	Apr. 1 – May 1
Onion, seed	Mar. 1 – Apr. 1	Mar. 15 – Apr. 1	Mar. 15 – Apr. 15
Onion, sets	Mar. 1 – Apr. 1	Mar 10 – Apr. 1	Mar. 10 – Apr. 10
Parsley	Mar. 10 – Apr. 10	Mar. 20 – Apr. 20	Apr. 1 – May 1
Parsnip	Mar. 10 – Apr. 10	Mar. 20 – Apr. 20	Apr. 1 – May 1
Peas, garden	Feb. 20 – Mar. 20	Mar. 10 – Apr. 10	Mar. 20 – May 1
Peas, black-eye	May 1 – July 1	May 10 – June 15	May 15 – June 1
Pepper [1]	May 1 – June 1	May 10 – June 1	May 15 – June 10
Potato	Mar. 10 – Apr. 1	Mar. 15 – Apr. 10	Mar. 20 – May 10
Radish	Mar. 1 – May 1	Mar. 10 – May 10	Mar. 20 – May 10
Rhubarb [1]	Mar. 1 – Apr. 1	Mar. 10 – Apr. 10	Mar. 20 – Apr. 15
Rutabaga			May 1 – June 1
Salsify	Mar. 10 – Apr. 15	Mar. 20 – May 1	Apr. 1 – May 1
Shallot	Mar. 1 – Apr. 1	Mar. 15 – Apr. 15	Apr. 1 – May 1
Sorrel	Mar. 1 – Apr. 15	Mar. 15 – May 1	Apr. 1 – May 15
Soybean	May 1 – June 30	May 10 – June 20	May 15 – June 15
Spinach	Feb. 15 – Apr. 1	Mar. 1 – Apr. 15	Mar. 20 – Apr. 20
Spinach, New Zealand	Apr. 20 – June 1	May 1 – June 15	May 1 – June 15
Squash, summer	Apr. 20 – June 1	May 1 – June 15	May 1 – 30
Sweet potato	May 1 – June 1	May 10 – June 10	May 20 – June 10
Tomato	Apr. 20 – June 1	May 5 – June 10	May 10 – June 15
Turnip	Mar. 1 – Apr. 1	Mar. 10 – Apr. 1	Mar. 20 – May 1
Watermelon	Apr. 20 – June 1	May 1 – June 15	May 15 – June 15

[1] Plants
[2] Generally fall planted

Planting dates for localities in which average date of last freeze is—

May 10	May 20	May 30	June 10
Mar. 10 – Apr. 30 _____	Apr. 20 – May 15 _____	May 1 – June 1 _____	May 15 – June 1.
May 25 – June 15 _____			
May 10 – June 30 _____	May 15 – June 20 _____	May 25 – June 15 _____	
Apr. 15 – June 15 _____	Apr. 25 – June 15 _____	May 1 – June 15 _____	May 15 – June 15.
Apr. 15 – June 1 _____	May 1 – June 15 _____	May 10 – June 10 _____	May 20 – June 10.
Apr. 15 – June 1 _____	May 1 – June 15 _____	May 10 – June 10 _____	May 20 – June 10.
Apr. 1 – May 15 _____	May 1 – June 15 _____	May 10 – June 15 _____	May 20 – June 1.
Apr. 1 – May 15 _____	May 1 – June 15 _____	May 10 – June 15 _____	May 20 – June 1.
Apr. 20 – June 15 _____	May 1 – June 1 _____	May 10 – June 1 _____	May 20 – June 1.
Apr. 15 – May 15 _____	May 10 – June 15 _____	May 20 – June 1 _____	June 1 – June 15.
Apr. 20 – June 15 _____	May 10 – June 15 _____	May 20 – June 1 _____	June 1 – June 15.
Apr. 20 – June 15 _____	May 10 – June 15 _____	May 20 – June 1 _____	June 1 – June 15.
Apr. 1 – May 1 _____	Apr. 15 – May 15 _____	May 1 – June 1 _____	May 15 – June 1.
June 1 – 20 _____	June 1 – 15 _____	June 1 – 15 _____	June 1 – 15.
Apr. 15 – June 1 _____	May 1 – June 1 _____	May 10 – June 1 _____	May 20 – June 1.
Apr. 1 – June 1 _____	Apr. 15 – June 1 _____	May 1 – June 15 _____	May 15 – June 15.
May 10 – June 1 _____	May 15 – June 1 _____	May 20 – June 1 _____	
Apr. 20 – May 20 _____	May 1 – June 1 _____	May 15 – June 1 _____	May 15 – June 15.
May 20 – June 15 _____	June 1 – 15 _____		
May 20 – June 15 _____	June 1 – 15 _____		
Apr. 15 – May 15 _____	May 1 – 30 _____	May 1 – 30 _____	May 15 – June 1.
Apr. 15 – May 15 _____	May 1 – 30 _____	May 1 – 30 _____	May 15 – June 1.
Apr. 1 – May 1 _____	Apr. 15 – May 15 _____	May 1 – 30 _____	May 15 – June 1.
Apr. 15 – May 15 _____	Apr. 20 – May 20 _____	May 1 – 30 _____	May 15 – June 1.
Apr. 10 – May 1 _____	Apr. 20 – May 10 _____	May 1 – 30 _____	May 15 – June 1.
Apr. 10 – May 15 _____	Apr. 20 – May 20 _____	May 1 – 30 _____	May 15 – June 1.
Apr. 15 – May 15 _____	May 1 – May 20 _____	May 1 – 15 _____	May 1 – 15.
Apr. 15 – May 15 _____	May 1 – June 30 _____	May 10 – June 30 _____	May 20 – June 30.
Apr. 15 – June 15 _____	May 1 – June 30 _____	May 10 – June 30 _____	May 20 – June 30.
June 1 – June 15 _____			
Apr. 15 – June 1 _____	May 1 – June 30 _____	May 10 – June 30 _____	May 20 – June 30.
May 20 – June 10 _____	June 1 – 20 _____		
Apr. 10 – May 1 _____	Apr. 20 – May 15 _____	May 1 – 30 _____	May 10 – June 10.
Apr. 1 – May 1 _____	Apr. 20 – May 15 _____	May 1 – 30 _____	May 10 – June 10.
Apr. 10 – May 1 _____	Apr. 20 – May 15 _____	May 1 – 30 _____	May 10 – June 10.
Apr. 15 – May 15 _____	May 1 – 20 _____	May 10 – June 1 _____	May 20 – June 10.
Apr. 15 – May 15 _____	May 1 – 20 _____	May 10 – June 1 _____	May 20 – June 10.
Apr. 1 – May 15 _____	Apr. 15 – June 1 _____	May 1 – June 15 _____	May 10 – June 15.
May 20 – June 10 _____	May 25 – June 15 _____	June 1 – 15 _____	
Apr. 1 – June 1 _____	Apr. 15 – June 15 _____	May 1 – June 15 _____	May 15 – June 1.
Apr. 1 – June 1 _____	Apr. 15 – June 15 _____	May 1 – June 15 _____	May 15 – June 1.
Apr. 1 – May 1 _____	Apr. 15 – May 10 _____	May 1 – 20 _____	May 15 – June 1.
May 1 – June 1 _____	May 1 – 20 _____	May 10 – 20 _____	May 20 – June 1.
Apr. 15 – June 1 _____	May 1 – June 1 _____	May 10 – June 1 _____	May 20 – June 1.
Apr. 10 – May 1 _____	Apr. 20 – May 10 _____	May 1 – June 1 _____	May 10 – June 1.
Apr. 15 – June 1 _____	May 1 – June 1 _____	May 10 – June 10 _____	May 20 – June 10.
May 25 – June 10 _____			
Apr. 1 – June 15 _____	Apr. 10 – June 15 _____	Apr. 20 – June 15 _____	May 1 – June 15.
May 10 – June 15 _____	May 20 – June 15 _____	June 1 – 15 _____	
May 10 – June 10 _____	May 20 – June 15 _____	June 1 – 20 _____	June 10 – 20.
May 15 – June 10 _____	May 25 – June 15 _____	June 5 – 20 _____	June 15 – 30.
Apr. 1 – June 1 _____	Apr. 15 – June 1 _____	May 1 – June 15 _____	May 15 – June 15.
June 1 – June 15 _____	June 15 – July 1 _____		

Source: United States Department of Agriculture

FALL PLANTING GUIDE FOR VEGETABLES AND HERBS

Crop	Planting dates for localities in which average date of last freeze is—		
	August 30	September 10	September 20
Asparagus [1]			
Beans, lima			
Beans, snap		May 15 – June 15	June 1 – July 1
Beet	May 15 – June 15	May 15 – June 15	June 1 – July 1
Broccoli, sprouting	May 1 – June 1	May 1 – June 1	May 1 – June 15
Brussels sprouts [1]	May 1 – June 1	May 1 – June 1	May 1 – June 15
Cabbage [1]	May 1 – June 1	May 1 – June 1	May 1 – June 15
Cabbage, Chinese	May 15 – June 15	May 15 – June 15	June 1 – July 1
Carrot	May 15 – June 15	May 15 – June 15	June 1 – July 1
Cauliflower [1]	May 1 – June 1	May 1 – July 1	May 1 – July 1
Celery [1] and celeriac	May 1 – June 1	May 15 – June 15	May 15 – July 1
Chard	May 15 – June 15	May 15 – July 1	June 1 – July 1
Chervil and chives	May 10 – June 10	May 1 – June 15	May 15 – June 15
Chicory, witloof	May 15 – June 15	May 15 – June 15	May 15 – June 15
Collards [1]	May 15 – June 15	May 15 – June 15	May 15 – June 15
Cornsalad	May 15 – June 15	May 15 – July 1	June 15 – Aug.1
Corn, sweet			June 1 – July 1
Cress, upland	May 15 – June 15	May 15 – July 1	June 15 – Aug. 1
Cucumber			June 1 – 15
Eggplant [1]			
Endive	June 1 – July 1	June 1 – July 1	June 15 – July 15
Fennel, Florence	May 15 – June 15	May 15 – July 15	June 1 – July 1
Garlic	(2)	(2)	(2)
Horseradish [1]	(2)	(2)	(2)
Kale	May 15 – June 15	May 15 – June 15	June 1 – July 1
Kohlrabi	May 15 – June 15	June 1 – July 1	June 1 – July 15
Leek	May 1 – June 1	May 1 – June 1	(2)
Lettuce, head [1]	May 15 – July 1	May 15 – July 1	June 1 – July 15
Lettuce, leaf	May 15 – July 15	May 15 – July 15	June 1 – Aug. 1
Muskmelon			May 1 – June 15
Mustard	May 15 – July 15	May 15 – July 15	June 1 – Aug. 1
Okra			June 1 – 20
Onion [1]	May 1 – June 10	May 1 – June 10	(2)
Onion, seed	May 1 – June 1	May 1 – June 10	(2)
Onion, sets	May 1 – June 1	May 1 – June 10	(2)
Parsley	May 15 – June 15	May 1 – June 15	June 1 – July 1
Parsnip	May 15 – June 1	May 1 – June 15	May 15 – June 15
Peas, garden	May 10 – June 15	May 1 – July 1	June 1 – July 15
Peas, black-eye			
Pepper [1]			June 1 – June 20
Potato	May 15 – June 1	May 1 – June 15	May 1 – June 15
Radish	May 1 – July 15	May 1 – Aug. 1	June 1 – Aug. 15
Rhubarb [1]	Sept. 1 – Oct. 1	Sept. 15 – Oct. 15	Sept. 15 – Nov. 1
Rutabaga	May 15 – June 15	May 1 – June 15	June 1 – July 1
Salsify	May 15 – June 1	May 10 – June 10	May 20 – June 20
Shallot	(2)	(2)	(2)
Sorrel	May 15 – June 15	May 1 – June 15	June 1– July 1
Soybean			
Spinach	May 15 – July 1	May 15 – July 1	June 1 – Aug. 1
Spinach, New Zealand			
Squash, summer	June 10 – 20	June 1 – 20	May 15 – July 1
Squash, winter			May 20 – June 10
Sweet potato			
Tomato	June 20 – 30	June 10 – 20	June 1 – 20
Turnip	May 15 – June 15	June 1 – July 1	June 1 – July 15
Watermelon			May 1 – June 15

[1] Plants
[2] Generally fall planted

Planting dates for localities in which average date of last freeze is—

September 30	October 10	October 20
	Oct. 20 – Nov. 15	Nov. 1 – Dec. 15.
June 1 – 15	June 1 – 15	June 15 – 30.
June 1 – July 10	June 15 – July 20	July 1 – Aug. 1.
June 1 – July 10	June 15 – July 25	July 1 – Aug. 5.
June 1 – 30	June 15 – July 15	July 1 – Aug. 1.
June 1 – 30	June 15 – July 15	July 1 – Aug. 1.
June 1 – July 10	June 1 – July 15	July 1 – 20.
June 1 – July 15	June 15 – Aug. 1	July 15 – Aug. 15.
June 1 – July 10	June 1 – July 20	June 15 – Aug. 1.
May 10 – July 15	June 1 – July 25	July 1 – Aug. 5.
June 1 – July 5	June 1 – July 15	June 1 – Aug. 1.
June 1 – July 5	June 1 – July 20	June 1 – Aug. 1.
(2)	(2)	(2)
June 1 – July 1	June 1 – July 1	June 15 – July 15.
June 15 – July 15	July 1 – Aug. 1	July 15 – Aug. 15.
July 15 – Sept. 1	Aug. 15 – Sept. 15	Sept. 1 – Oct. 15.
June 1 – July 1	June 1 – July 10	June 1 – July 20.
July 15 – Sept. 1	Aug. 15 – Sept. 15	Sept. 1 – Oct. 15.
June 1 – July 1	June 1 – July 1	June 1 – July 15.
May 20 – June 10	May 15 – June 15	June 1 – July 1.
June 15 – Aug. 1	July 1 – Aug. 15	July 15 – Sept. 1.
June 1 – July 1	June 15 – July 15	June 15 – Aug. 1.
(2)	(2)	(2)
(2)	(2)	(2)
June 15 – July 15	July 1 – Aug. 1	July 15 – Aug. 15.
June 15 – July 15	July 1 – Aug. 1	July 15 – Aug. 15.
(2)	(2)	(2)
June 15 – Aug. 1	July 15 – Aug. 15	Aug. 1 – 30.
June 1 – Aug. 1	July 15 – Sept. 1	July 15 – Sept. 1.
May 15 – June 1	June 1 – June 15	June 15 – July 20.
June 15 – Aug. 1	July 15 – Aug. 15	Aug. 1 – Sept. 1.
June 1 – July 1	June 1 – July 15	June 1 – Aug. 1.
(2)	(2)	(2)
(2)	(2)	(2)
(2)	(2)	(2)
June 1 – July 15	June 15 – Aug. 1	July 15 – Aug. 15.
June 1 – July 1	June 1 – July 10	(2)
June 1 – Aug. 1	(2)	(2)
----	June 1 – July 1	June 1 – July 1.
June 1 – July 1	June 1 – July 1	June 1 – July 10.
May 1 – June 15	May 15 – June 15	June 15 – July 15.
July 1 – Sept. 1	July 15 – Sept. 15	Aug. 1 – Oct. 1.
Oct. 1 – Nov. 1	Oct. 15 – Nov. 15	Oct. 15 – Dec. 1.
June 1 – July 1	June 15 – July 15	July 10 – 20.
June 1 – 20	June 1 – July 1	June 1 – July 1.
(2)	(2)	(2)
June 1 – July 15	July 1 – Aug. 1	July 15 – Aug. 15.
May 25 – June 10	June 1 – 25	June 1 – July 5.
July 1 – Aug. 15	Aug. 1 – Sept. 1	Aug. 20 – Sept. 10.
May 15 – July 1	June 1 – July 15	June 1 – Aug. 1.
June 1 – July 1	June 1 – July 15	June 1 – July 20.
June 1 – 15	June 1 – July 1	June 1 – July 1.
----	May 20 – June 10	June 1 – 15.
June 1 – 20	June 1 – 20	June 1 – July 1.
June 1 – Aug. 1	July 1 – Aug. 1	July 15 – Aug. 15.
May 15 – June 1	June 1 – June 15	June 15 – July 20.

Crop	Planting dates for localities in which average date of last freeze is—		
	Oct. 30	Nov. 10	Nov. 20
Asparagus [1]	Nov. 15 – Jan. 1	Dec. 1 – Jan. 1	
Beans, lima	July 1 – Aug. 1	July 1 – Aug. 15	July 15 – Sept. 1
Beans, snap	July 1 – Aug. 15	July 1 – Sept. 1	July 1 – Sept. 10
Beet	Aug. 1 – Sept. 1	Aug. 1 – Oct. 1	Sept. 1 – Dec. 1
Broccoli, sprouting [1]	July 1 – Aug. 15	Aug. 1 – Sept. 1	Aug. 1 – Sept. 15
Brussel sprouts [1]	July 1 – Aug. 15	Aug. 1 – Sept. 1	Aug. 1 – Sept. 15
Cabbage [1]	Aug. 1 – Sept. 1	Sept. 1 – 15	Sept. 1 – Dec. 1
Cabbage, Chinese	Aug. 1 – Sept. 15	Aug. 15 – Oct. 1	Sept. 1 – Oct. 15
Carrot	July 1 – Aug. 15	Aug. 1 – Sept. 1	Sept. 1 – Nov. 1
Cauliflower [1]	July 15 – Aug. 15	Aug. 1 – Sept. 1	Aug. 1 – Sept. 15
Celery and celeriac	June 15 – Aug. 15	July 1 – Aug. 15	July 15 – Sept. 1
Chard	June 1 – Sept. 10	June 1 – Sept. 15	June 1 – Oct. 1
Chervil and chives	(2)		Nov. 1 – Dec. 31
Chicory, witloof	July 1 – Aug. 10	July 10 – Aug. 20	July 20 – Sept. 1
Collards [1]	Aug. 1 – Sept. 15	Aug. 15 – Oct. 1	Aug. 25 – Nov. 1
Cornsalad	Sept. 15 – Nov. 1	Oct. 1 – Dec. 1	Oct. 1 – Dec. 1
Corn, sweet	June 1 – Aug. 1	June 1 – Aug. 15	June 1 – Sept. 1
Cress, upland	Sept. 15 – Nov. 1	Oct. 1 – Dec. 1	Oct. 1 – Dec. 1
Cucumber	June 1 – Aug. 1	June 1 – Aug. 15	June 1 – Aug. 15
Eggplant [1]	June 1 – July 1	June 1 – July 15	June 1 – Aug. 1
Endive	July 15 – Aug. 15	Aug. 1 – Sept. 1	Sept. 1 – Oct. 1
Fennel, Florence	July 1 – Aug. 1	July 15 – Aug. 15	Aug. 15 – Sept.15
Garlic	(2)	Aug. 1 – Oct. 1	Aug. 15 – Oct. 1
Horseradish [1]	(2)	(2)	(2)
Kale	July 15 – Sept. 1	Aug. 1 – Sept. 15	Aug. 15 – Oct. 15
Kohlrabi	Aug. 1 – Sept. 1	Aug. 15 – Sept. 15	Sept. 1 – Oct. 15
Leek	(2)	(2)	Sept. 1 – Nov. 1
Lettuce, head [1]	Aug. 1 – Sept. 15	Aug. 15 – Oct. 15	Sept. 1 – Nov. 1
Lettuce, leaf	Aug. 15 – Oct. 1	Aug. 25 – Oct. 1	Sept. 1 – Nov. 1
Muskmelon	July 1 – July 15	July 15 – July 30	
Mustard	Aug. 15 – Oct. 15	Aug. 15 – Nov. 1	Sept. 1 – Dec. 1
Okra	June 1 – Aug. 10	June 1 – Aug. 20	June 1 – Sept. 10
Onion [1]		Sept. 1 – Oct. 15	Oct. 1 – Dec. 31
Onion, seed			Sept. 1 – Nov. 1
Onion, sets		Oct. 1 – Dec. 1	Nov. 1 – Dec. 31
Parsley	Aug. 1 – Sept. 15	Sept. 1 – Nov. 15	Sept. 1 – Dec. 31
Parsnip	(2)	(2)	Aug. 1 – Sept. 1
Peas, garden	Aug. 1 – Sept. 15	Sept. 1 – Nov. 1	Oct. 1 – Dec. 1
Peas, black-eye	June 1 – Aug. 1	June 15 – Aug. 15	July 1 – Sept. 1
Pepper [1]	June 1 – July 20	June 1 – Aug. 1	June 1 – Aug. 15
Potato	July 20 – Aug. 10	July 25 – Aug. 20	Aug. 10 – Sept. 15
Radish	Aug. 15 – Oct. 15	Sept. 1 – Nov. 15	Sept. 1 – Dec. 1
Rhubarb [1]	Nov. 1 – Dec. 1		
Rutabaga	July 15 – Aug. 1	June 15 – Aug. 15	Aug. 1 – Sept. 1
Salsify	June 1 – July 10	June 15 – July 20	July 15 – Aug. 15
Shallot	(2)	Aug. 1 – Oct. 1	Aug. 15 – Oct. 1
Sorrel	Aug. 1 – Sept. 15	Aug. 15 – Oct. 1	Aug. 15 – Oct. 15
Soybean	June 1 – July 15	June 1 – July 25	June 1 – July 20
Spinach	Sept. 1 – Oct. 1	Sept. 15 – Nov. 1	Oct. 1 – Dec. 1
Spinach, New Zealand	June 1 – Aug. 1	June 1 – Aug. 15	June 1 – Aug. 15
Squash, summer	June 1 – Aug. 1	June 1 – Aug. 10	June 1 – Aug. 20
Squash, winter	June 10 – July 10	June 20 – July 20	July 1 – Aug. 1
Sweet potato	June 1 – 15	June 1 – July 1	June 1 – July 1
Tomato	June 1 – July 1	June 1 – July 15	June 1 – Aug. 1
Turnip	Aug. 1 – Sept. 15	Sept. 1 – Oct. 15	Sept. 1 – Nov. 15
Watermelon	July 1 – July 15	July 15 – July 30	

[1] Plants

[2] Generally fall planted

Planting dates for localities in which average date of last freeze is—

Nov. 30	Dec. 10	Dec. 20
Aug. 1 – 15	Sept. 1 – 30	
Aug. 15 – Sept. 20	Sept. 1 – 30	
Sept. 1 – Dec. 15	Sept. 1 – Dec. 31	Sept. 1 – Oct. 1.
Aug. 1 – Oct. 1	Aug. 1 – Nov. 1	Sept. 1 – Nov. 1.
Aug. 1 – Oct. 1	Aug. 1 – Nov. 1	Sept. 1 – Dec. 31.
Sept. 1 – Dec. 31	Sept. 1 – Dec. 31	Sept. 1 – Dec. 31.
Sept. 1 – Nov. 1	Sept. 1 – Nov. 15	Sept. 1 – Dec. 31.
Sept. 15 – Dec. 1	Sept. 15 – Dec. 1	Sept. 1 – Dec. 31.
Aug. 15 – Oct. 10	Sept. 1 – Oct. 20	Sept. 1 – Dec. 1.
Aug. 1 – Dec. 1	Sept. 1 – Dec. 31	Sept. 15 – Dec. 1.
June 1 – Nov. 1	June 1 – Dec. 1	Sept 15 – Nov. 1.
Nov. 1 – Dec. 31	Nov. 1 – Dec. 31	Oct. 1 – Dec. 31.
Aug. 15 – Sept. 30	Aug. 15 – Oct. 15	June 1 – Dec. 31.
Sept. 1 – Dec. 1	Sept. 1 – Dec. 31	Nov. 1 – Dec. 31.
Oct. 1 – Dec. 31	Oct. 1 – Dec. 31	Aug. 15 – Oct. 15.
		Sept. 1 – Dec. 31.
		Oct. 1 – Dec. 31.
Oct. 1 – Dec. 31	Oct. 1 – Dec. 31	Oct. 1 – Dec. 31.
July 15 – Sept. 15	Aug. 15 – Oct. 1	Aug. 15 – Oct. 1.
July 1 – Sept. 1	Aug. 1 – Sept. 30	Aug. 1 – Sept. 30.
Sept. 1 – Nov. 15	Sept. 1 – Dec. 31	Sept. 1 – Dec. 31.
Sept. 1 – Nov. 15	Sept. 1 – Dec. 1	Sept. 1 – Dec. 1.
Sept. 1 – Nov. 15	Sept. 15 – Nov. 15	Sept. 15 – Nov. 15.
(2)	(2)	(2)
Sept. 1 – Dec. 1	Sept. 1 – Dec. 31	Sept. 1 – Dec. 31.
Sept. 1 – Dec. 1	Sept. 15 – Dec. 31	Sept. 1 – Dec. 31.
Sept. 1 – Nov. 1	Sept. 1 – Nov. 1	Sept. 15 – Nov. 1.
Sept. 1 – Dec. 1	Sept. 15 – Dec. 31	Sept. 15 – Dec. 31.
Sept. 1 – Dec. 1	Sept. 15 – Dec. 31	Sept. 15 – Dec. 31.
Sept. 1 – Dec. 1	Sept. 1 – Dec. 1	Sept. 15 – Dec. 1.
June 1 – Sept. 20	Aug. 1 – Oct. 1	Aug. 1 – Oct. 1.
Oct. 1 – Dec. 31	Oct. 1 – Dec. 31	Oct. 1 – Dec. 31.
Sept. 1 – Nov. 1	Sept. 1 – Nov. 1	Sept. 1 – Nov. 1.
Nov. 1 – Dec. 31	Nov. 1 – Dec. 31	Nov. 1 – Dec. 31.
Sept. 1 – Dec. 31	Sept. 1 – Dec. 31	Sept. 1 – Dec. 31.
Sept. 1 – Nov. 15	Sept. 1 – Dec. 1	Sept. 1 – Dec. 1.
Oct. 1 – Dec. 31	Oct. 1 – Dec. 31	Oct. 1 – Dec. 31.
July 1 – Sept. 10	July 1 – Sept. 20	July 1 – Sept. 20.
June 15 – Sept. 1	Aug. 15 – Oct. 1	Aug. 15 – Oct. 1.
Aug. 1 – Sept. 15	Aug. 1 – Sept. 15	Aug. 1 – Sept. 15.
Sept. 1 – Dec. 31	Aug. 1 – Sept. 15	Oct. 1 – Dec. 31.
Sept. 1 – Nov. 15	Oct. 1 – Nov. 15	Oct. 15 – Nov. 15.
Aug. 15 – Sept. 30	Aug. 15 – Oct. 15	Sept. 1 – Oct. 31.
Aug. 15 – Oct. 15	Sept. 15 – Nov. 1	Sept. 15 – Nov. 1.
Sept. 1 – Nov. 15	Sept. 1 – Dec. 15	Sept. 1 – Dec. 31.
June 1 – July 30	June 1 – July 30	June 1 – July 30.
Oct. 1 – Dec. 31	Oct. 1 – Dec. 31	Oct. 1 – Dec. 31.
June 1 – Sept. 1	June 1 – Sept. 15	June 1 – Oct. 1.
July 15 – Aug. 15	Aug. 1 – Sept. 1	Aug. 1 – Sept. 1.
June 1 – July 1	June 1 – July 1	June 1 – July 1.
Aug. 1 – Sept. 1	Aug. 15 – Oct. 1	Sept. 1 – Nov. 1.
Sept. 1 – Nov. 15	Oct. 1 – Dec. 1	Oct. 1 – Dec. 31.

Source: United States Department of Agriculture

USDA HARDINESS ZONE MAP

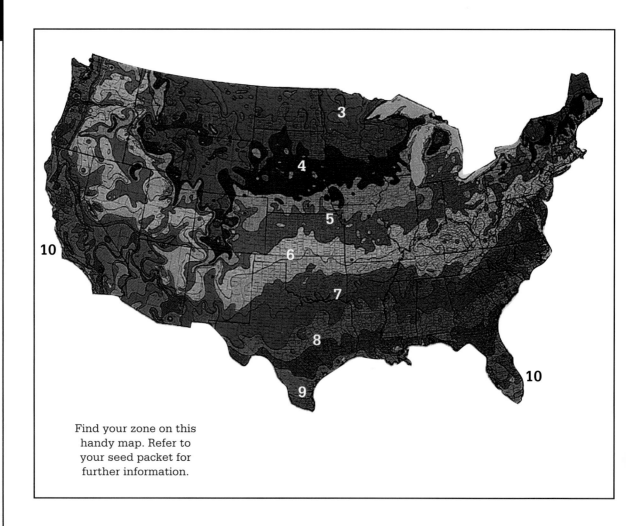

Find your zone on this handy map. Refer to your seed packet for further information.

SEEDS THAT NEED LIGHT TO GERMINATE

FLOWERS

Achillea—yarrow, milfoil

Ageratum Houstonianum—ageratum, floss flower

Antirrhinus majus—snapdragon

Aquilegia—columbine

Begonia—begonia

Brassica oleracea Acephala Group—flowering kale, flowering cabbage

Browallia speciosa—browallia

Campanula—bellflower, harebell, Canterbury bells

Chyrsanthemum parthenium—feverfew, matricaria

Chrysanthemum × superbum (C. maximum)—shasta daisy

Coleus × hybridus—coleus

Coreopsis grandiflora—coreopsis, tickseed

Cortaderia selloana—pampas grass

Cuphea ignea—cigar flower, cigar plant, firecracker plant

Doronicum cordatum—leopard's-bane

Fuchsia—fuchsia, lady's eardrops

Gaillardia—blanketflower

Gerbera Jamesonii—gerbera, Transvaal daisy, African daisy

Helichrysum bractetum—strawflower, immortelle

Hesperis matronalis—dame's rocket, sweet rocket, garden rocket

Impatiens—impatiens, busy Lizzy, balsam, touch-me-not

Lobularia maritima—sweet alyssum

Lychnis chalcedonica—Maltese cross, campion, Jerusalem cross

Matthiola incana—stock, gillyflower

Molucella laevis—bells of Ireland

Nicotiana alata—flowering tobacco

Papaver orientale—Oriental poppy

Perilla frutescens—perilla, false coleus, beefsteak plant

Petunia × *hybrida*—petunia

Physalis Alkekengi—Chinese lanterns, winter cherry, bladder cherry

Platycodon grandiflorus—balloon flower, Chinese bellflower

Primula × *polyantha*—primrose

Reseda odorata—mignonette

Salvia—salvia

Sanvitalia produmbens—creeping zinnia, trailing sanvitalia

Tithonia rotundifolia—Mexican sunflower

Venidium fastuosum—Cape daisy, monarch of the veldt

HERBS AND VEGETABLES

Anethum graveolens—dill

Capsicum annuun var. *annuum*—pepper

Lactuca sativa—lettuce

Satureja—savory

INFORMATION SOURCES

INFORMATION AVAILABLE FROM THE BURPEE COMPANY

To order a Burpee catalog, call, write, or fax:

W. Atlee Burpee & Co.
300 Park Avenue
Warminster, PA 18974
Telephone: 800-888-1447
Fax: 800-487-5530

Burpee also publishes *Burpee Home Gardener*, an excellent source for reliable and practical information on and solutions to a variety of gardening topics from seed starting to garden design. The magazine is published four times a year and is available at newsstands or by subscription. To order a subscription, call 800-888-1447.

You can contact Burpee for a catalog and to get information about seed starting and new plant varieties by visiting Burpee's World Wide Web home page. The Web address is http://garden.burpee.com

ADDITIONAL INFORMATION SOURCES

County Extension Agencies

Your local county extension agency is available to help with your gardening questions and problems and to conduct soil testing. Extension agencies are usually listed in the blue pages, or government services section, of your phone book.

Botanical Gardens and Arboreta

Many botanical and public gardens have educational classes, libraries, and telephone hotline information services. Call your nearest public garden and ask about the services that they provide concerning gardening information and assistance.

American Horticultural Society

The American Horticultural Society has the Gardener's Information Service telephone hotline to answer your gardening questions. Professional horticulturists are available

Monday through Friday from 11:00 a.m.–3:00 p.m. You can also write to them with questions at:

American Horticultural Society
Gardener's Information Service
7931 East Boulevard
Alexandria, VA 22308

Internet, Web, and other Online Information Sites

Gardening information sites are exploding on the Internet, World Wide Web, and commercial online services. They are excellent places to get information or to post a question to obtain answers from experts to fellow home gardeners. Here are a few sites with links to many other gardening sites on the Internet and Web.

Agropolis, the Texas A&M University System Agriculture Program:
http://agcomwww.tama.edu/agcom/agrotext/visitor.html

Garden Gate on Prairienet:
http://www.prairienet.org/ag/garden/homepage.htm

GardenNet:
http://www.olympus.net/gardens/welcome.htlm

Ohio State University's WebGarden:
http://hortwww-2.ag.ohio-state.edu/hvp/webgarden/webgarden.html

America Online:
AOL has a good gardening site. I like the message board sites for posting questions. You can usually count on getting some useful answers and advice by everyone from experts to experienced home gardeners within a few days' time or less.

GLOSSARY

Acidic: A pH of less than 7. The lower the number the more acidic.

Alkaline: A pH greater than 7. The higher the number the more alkaline.

Annual: A plant that lives for one year or one growing season.

Biennial: A plant that lives for two years or two growing seasons. These plants produce flowers and seed in the second year. Biennials often self-sow the second year and thus appear to act as perennials, which come back every year.

Compost: Organic materials (grass clippings, leaves, manures, vegetable scraps, etc.) that have decomposed into a crumbly soil like material that is added to garden beds as a soil amendment and/or mulch.

Cotyledon: The embryonic leaves that are first visible from a germinated seed. They are also called the seed leaves and wither off once the first true leaves develop.

Cultivar: A contraction of the term *cultivated variety*. A cultivated variety is a plant species that can be differentiated in some way from the straight species. The new variety or distinguishing feature(s) can be reproduced by seed or vegetative propagation.

Cutting back: Shearing back plants, usually perennials, after the first flowering period to encourage a second one and/or to obtain a neater appearance.

Damping off: A soil fungal disease that collapses the stems of seedlings. Usually caused by over- watering, poor air circulation, overcrowding, or unsterile soil.

Double flowers: Blooms with two or more rows of petals.

Full sun: A garden site that receives a minimum of 6 hours of strong, direct sunlight daily.

Fungicide: A chemical formulation (usually a spray or powder) that helps prevent or control soil or plant fungal diseases.

Germination: When a seed begins to grow and develop into a plant.

Hardening off: The process of gradually exposing indoor grown plants to outdoor growing conditions.

Humus: Decayed plant and animal material, also referred to as organic matter, that is added to soil to improve drainage, structure, and fertility.

Hybrid seed: Seed produced from cross-fertilizing two different plant varieties. The first generation of plants of the resulting seed is referred to as the F1 hybrids. Seed collected and grown from F1 hybrids are called the F2 hybrids (second generation); plants grown from F2 seed will not come true to type (will not have the same characteristics as the F1 hybrids).

Inflorescence: The arrangement of flowers on a plant or the shape of the flowering part of the plant.

Leaf mold: Partly decomposed leaves.

Loam: A soil type that is a combination of sand, silt, and about 25 percent clay.

N-P-K: Nitrogen-phosphorus-potassium. These are the three major, or macro, nutrients plants need for proper and healthy development.

Open pollinated: Refers to nonhybrid plants or seeds. Seeds collected from open-pollinated plants will grow true to type (will display the same characteristics as the parent plants).

Organic matter: Plant and animal matter in various stages of decomposition. Humus, compost, and leaf mold are often referred to as organic matter.

Part shade: A garden site that receives 2–6 hours of direct sun daily.

Peat moss: Decomposed materials made from bog mosses and sedges. Added to soil mixes to increase moisture retention and drainage.

Perennial: A plant that produces flowers, seeds, and/or fruit every year. Herbaceous perennials die back during the winter and grow again in the spring. Woody perennials develop woody stems and while some may lose their leaves during the fall or winter, they do not die back.

Perlite: Volcanic rock particles that have been superheated. Used to improve drainage and air circulation in soil mixes.

pH: A scale used to measure acidity or alkalinity, which ranges from 0 to 14. A value of 7 indicates neutrality; higher values indicate alkalinity and lower, acidity.

Pinch back: Snipping off the growing tips or flower buds of plants to encourage fuller or bushier growth.

Pot-bound: When a plant's roots become matted and snarled together because the growing container has become too small for healthy root growth.

Raceme: An erect, elongated inflorescence that bears flowers close along the flowering stem.

Rhizome: An underground stem that has either nodes, buds, or scalelike leaves.

Rust: A fungal disease that causes reddish brown discoloration and spotting of foliage. It damages foliage and can stunt a plant's growth.

Scarification: Cutting or scratching thick seed coats to help the seed absorb water, which speeds up germination.

Shade: A garden site that receives little to no direct sunlight.

Single flowers: Blooms with just one row of petals.

Species: A category of plants or animals. All plants have a botan-ical, or Latin, name with at least two parts. The first name is the genus name, which usually includes a large group of plants that share a number of general characteristics.

The second name is the species name, which is a subdivision of the genus. Plants of the same species share many specific characteristics.

Stratification: The process of giv-ing seeds a cold treatment before sow-ing. Usually, seeds are placed in moistened medium and exposed to 32–40°F for a period of time, which varies according to variety.

Succulent: Refers to thick, usually fleshy, juicy leaves.

Thinning: Removing some seedlings or plants so the remain-ing ones have adequate room for healthy growth and develoment.

Tubular: Refers to a funnel- or trumpet-type flower shape.

Umbel: A usually flat-topped inflo-rescence that resembles the stays of an umbrella. A Queen Anne's lace flower is a good example.

Vermiculite: Mica rock that has been heated to break it into very lightweight particles. Used in soil and soilless mixtures.

NEW FLOWERS FOR -1889-
GROWN BY
W. ATLEE BURPEE & CO. PHILADELPHIA & LONDON.

BIBLIOGRAPHY

Abraham, Doc, and Katy Abraham. *Growing Plants from Seed.* New York: Lyons & Burford, 1991.

Armitage, Allan, Maureen Heffernan, Chela Kleiber, and Holly Schimizu. *Burpee Complete Gardener.* New York: Macmillan, 1995.

Brickell, Christopher, ed. *The American Horticultural Society Encyclopedia of Garden Plants.* New York: Macmillan, 1992.

Bubel, Nancy. *The New Seed Starter's Handbook.* Emmaus, Pa.: Rodale Press, 1988.

Butterfield, Bruce. "On Your Mark . . . Get Set . . . Start Your Seeds!" *National Gardening Magazine.* January/February 1996, pp. 46–49.

Editors of *Garden Way. Just the Facts!* Pownal, Vt.: Garden Storey Way, 1993.

Editors of *Mother Earth News. The Healthy Garden Handbook.* New York: Simon & Schuster, 1989.

Editors of Time-Life. *The Time-Life Complete Guide to Gardening & Landscaping.* New York: Prentice Hall, 1991.

Everett, Thomas. *New York Botanical Garden's Encyclopedia of Gardening.* 10 vols. New York: Garland, 1981.

Feinman, Jeffrey. *U.S. Gardening Guide.* New York: Fireside Books, 1979.

Fell, Derek. *Vegetables: How to Select, Grow, and Enjoy.* Los Angeles: HP Books, 1982.

Genders, Roy. *The Complete Book of Herbs and Herb Growing.* New York: Sterling, 1980.

Heffernan, Maureen. *Tips for Successsful Seed Germination.* Alexandria, Va.: American Horticultural Society, 1994.

Hill, Lewis, and Nancy Hill. *Successful Perennial Gardening: A Practical Guide.* Pownal, Vt.: Garden Storey Way, 1988.

Hively, Suzanne. "Plot Timing of Indoor Plantings." *Cleveland Plain Dealer.* January 5, 1996, pp. E1–E2.

Jeavons, John. *How to Grow More Vegetables.* Berkeley, Calif.: Ten Speed Press, 1982.

Lloyd, Christopher, and Graham Rice. *Garden Flowers from Seed.* Portland, Ore.: Timber Press, 1994.

Loewer, Peter. "How to Succeed Starting from Seed." *Burpee Home Gardener.* Winter 1996, pp. 18–21, 56–57.

Nau, Jim. *Ball Culture Guide: Encyclopedia of Seed Germination.* 2nd ed. Batavia, Il.: Ball, 1993.

Raver, Anne. "Fluorescent Light and Vivaldi Keep Seeds Happy." *New York Times.* February 18, 1996, p. 21.

Reilly, Ann. *Park's Success with Seeds.* Greenwood, S.C.: Geo. Park Seed Co., 1978.

Sax, Irene. "Cold Frames." *Martha Stewart Living.* March 1996, pp. 74–78.

"Sowing Support." *Practical Gardening.* January 1996, pp. 62–65.

Still, Steven. *Manual of Herbaceous Ornamental Plants.* Champaign, Il.: Stipes, 1988.

Swain, Roger. *The Practical Gardener.* New York; Henry Holt & Co., 1991.

White, Judy. "Annual Ritual." *National Gardening Magazine.* March–April, 1993, pp. 42–45.

Wyman's Gardening Encyclopedia. New York: Macmillan, 1986.

INDEX

Solomon's seal, 151
Sorrel, 216, *216,* 228–35
  French, 216, *216*
  garden, 216
Southern star, 142–43
Sowing
  frost-free dates for, 3, 5,
    220–22
  indoors versus outdoor, *4,*
    4–5
Soybean, 228–35
Spacing in outdoor seed
    sowing, 48–50
Spearmint, 212, 227
Sphagnum peat moss, 18,
    19
Spider flower, 89
Spiderwort, 165
Spike speedwell, 168
Spinach, 179, 195–96, 226,
    228–35
  New Zealand, 228–35
Spinacia oleracea, 195–96
  'Avon,' 196
  'Melody,' 196, *196*
Spirea, false, 73–74
Spring
  planting guide for,
    228–31
  sowing perennial needs in,
    55
Spring anemone, 153
Spurge, 102, *102*
Squash
  summer, 196–97, 226,
    228–35
  winter, 196–97, 226
Stachys byzantina, 161
Starflower, 224
Star of the Argentine,
    142–43
Star of the veldt, 97–98
Statice, 127
Stay-in-place, 149–50
Stock, 135, 237
  common, 135
Stokes' aster, 161–62
Stokesia laevis, 161–62
Stonecrop, 159–60
Strawberry, 197–98
Strawflower, 115, 237

String bean, 175–77
Sugar pea, 191–92
Summer, sowing perennial
    needs in, 56
Summer cypress, 123
Sundrop, Ozark, 142
Sunflower, 113–14, 226
  false, 113, 115–16
  Mexican, 164, 238
Sunflower heliopsis,
    115–16
Sun plant, 151
Sunrise coreopsis, 55
Sun rose, 113
Swamp rose, 118
Swan River daisy, 78, 104
Swedish turnip, 195
Sweet alyssum, 130, 237
Sweet marjoram, 211–12
Sweet pea, 124–25, 224
  annual, 124–25
Sweet potato, 228–35
Sweet rocket, 117, 237
Sweet William, 224
Swiss chard, 198, 226

T

Tagetes, 162–63
  'French Vanilla,' 162–63
  'Jaguar,' *162,* 163
  'Lady,' 163
  'Nugget Supreme Yellow,'
    163
Tahoka daisy, 133–34
Tanacetum vulgare,
    216–17
Tansy, 216–17
Tarragon, 217
  French, 217
Temperature in indoor seed
    germination, 12
Ten week stock, 135
Texas bluebonnet, 131
Texas pride, 147–48
Thalictruzm, 163
Thinning
  in indoor seed sowing, 34
  in outdoor seed sowing, 51

Thistle, 87–88
  globe, 100
  Japanese, 87–88
  plume, 87–88
Thrift, common, 71–72
Thunbergia alata, 163–64
  'Susie,' *163,* 164
Thyme, 217, 227
  common, 217
  culinary, 217
Thymus vulgaris, 217
Tickseed, 237
Tidytips, 125–26
Timing
  for starting annual seeds,
    2, 5
  for starting perennial
    seeds, 2
  for starting seeds out-
    doors, 45–46
Tithonia rotundifolia, 164,
    238
  'Torch,' 164
Tobacco, flowering, 140,
    237
Tomato, 198–200, 226,
    228–35
Torch lily, 122–23
Torenia fournieri, 164–65
  'Happy Faces,' *164,* 165
Touch-me-not, 120–21,
    121, 237
Trachymene coerulea, 165
  'Lacy,' 165
Tradescantia virginiana, 165
Trailing sanvitalia, 157,
    238
Transplanting in indoor
    seed germination,
    34–35
Transvaal daisy, 109, 237
Treasure flower, 108
Tree celandine, 134
Tree mallow, 125
Tritoma, 122–23
Tropaeolum majus, 166
  'Alaska,' *166*
Troubleshooting seed
    starting problems
  indoors, 38–42
  outdoors, 57

Tuberous begonia, 75–76
Turnip, 200, 226, 228–35
  Swedish, 195

U

USDA hardiness zone map,
    236

V

Valerian, Greek, 150–51
Vegetables, 174–201
  fall planting guide for,
    232–35
  indoor and outdoor plant-
    ing guide for, 225–26
  and need for light to ger-
    minate seeds, 237
  seed tapes for, 10
  spring planting guide for,
    228–31
Veldt daisy, 109
Venidium fastuosum,
    166–67, 238
Verbascum, 167
  'Southern Charm,' 167
Verbena
  annual, 167–68
  garden, 167–68
Verbena × Hybrida,
    167–68
  'Peaches & Cream,' 168,
    168
Vermiculite, 18–19
Veronica
  'Sightseeing,' 168
  spicata, 168
  Teucrium, 168
Vinca, 83–84, 224
Vinca rosea, 83–84
Viola, 224
Viola
  tricolor, 168–69, *169*
  × *Wittrockiana,* 169–70